Applying Maths
in Construction

Applying Maths in Construction

Student's book

Antoinette Tourret and John Humphreys
The University of Greenwich

A member of the Hodder Headline Group
LONDON • SYDNEY • AUCKLAND

Published in Great Britain in 1997 by Arnold,
a member of the Hodder Headline Group,
338 Euston Road, London NW1 3BH

British Library Cataloguing in Publication Data
A catalogue entry for this book is available from the British Library

ISBN 0 340 65295 0

Produced by Gray Publishing, Tunbridge Wells, Kent
Printed and bound in Great Britain by J W Arrowsmith Ltd, Bristol

Contents

Project 3

Project 4

Project 5

Introduction

This book has been designed to provide practice of the application of number skills within vocational courses in Construction and the Built Environment.

The ideas and techniques covered by this book and its accompanying *Teacher's Guide* have been chosen to suit all students taking courses such as:

- NVQ levels 1–3 in Construction
- GNVQ at Intermediate and Advanced levels in Construction and the Built Environment

Projects have been carefully selected to meet relevant needs only and have been presented entirely within a construction context.

About the projects

The projects provide an opportunity to see how number skills are applied when solving various problems within a range of construction contexts. Estimation is used to predict quantities and as a checking procedure for calculations. Wherever appropriate, checking procedures used in the workplace are incorporated into the projects. The projects also provide exemplars of appropriate evidence for the application of number skills in a vocational context.

Exercises are included at intervals during the development of the projects to provide an oportunity for practising the specific skills which have been used. Their completion will provide the student with appropriate evidence for the core skill of the application of number.

Each project introduction briefly describes the number skills used during its development and gives references to the topics and modules published in the *Teacher's Guide*. If further practice is required other than that given in the projects, the student will be able to use the modules from the resource pack to improve skills before tackling the exercises in the projects.

The sections within each of the projects are as follows:

A Number skills described in terms of topics and modules from the resource pack
B Job description and task analysis
C Materials cost
D Other considerations which include materials information and task costing.

Decorating a room

A Number skills

Addition	Modules 2, 4, 7
Algebra	Module 27
Approximating and estimating	Modules 2, 12, 14
Areas	Modules 15, 16
Checking procedures	Module 12
Converting units of measurement	Module 11
Division	Module 13
Fractions	Module 22
Measurement	Modules 1
Money	Modules 12, 13
Multiplication	Modules 12, 16
Ratios and proportions	Module 19
Percentages	Modules 4, 10
Perimeters	Modules 2, 7
Subtraction	Modules 2, 7
Technical drawing	Modules 5, 26

B Job description and analysis of tasks

MartinCrafts have been asked to provide an estimate for decorating a lounge which has been allowed to deteriorate badly due to lack of maintenance. At a preliminary visit, two of their decorators, Penny and David, took the following room dimensions and made a list of the tasks which would have to be carried out.

Room dimensions

- the room is rectangular – 6.9 m in length by 4.7 m in width
- the height is constant throughout the room at 2.45 m
- there are two doors (0.75 m by 1.95 m) at either end of the 6.9 m partition wall

- each door is 500 mm in from the corner of the room
- one window is situated on each 4.7 m wall of the lounge, measuring 1.05 m by 2.4 m
- the windows are positioned 1 m from the floor
- there is 50 mm architrave around each door and 50 mm cover fillets around each window
- the skirting board measures 125 mm in depth and runs round the whole room.

The preliminary visit done by Penny and David indicated that the walls and ceiling would require stripping completely and repairing where necessary. All paintwork will need to be burnt-off and reduced to the bare wood before re-painting. They drew up the following list:

- all walls to be stripped and sized
- all walls to be wall-papered
- the ceiling to be cleaned and textured
- 150 mm coving to be fitted
- the ceiling and coving to be given three coats of white emulsion
- the doors are to be stripped, primed, undercoated and glossed on one side to specification
- the windows are to be stripped, primed, undercoated and glossed on the inside
- all skirtings, architraves and coverfillets are to be stripped, primed, undercoated and glossed to specification.

C Cost of materials

Penny's and David's next task is to find the cost of materials from their suppliers. The prices given are without VAT added.

Emulsion	£25 per 5 litres (L)
Wallpaper	£5.45 per roll
Primer	£3.59 per L
Undercoat	£7.79 per 2.5 L
Gloss finish	£4.45 per L
Artex texture	£8.99 per 25 kg bag
Coving	£15.95 per 10 m pack
Size	£1.55 per packet

D Other considerations

After more meetings with the customer and considering the quality of the tasks with their manager, Penny and David make a list of important facts which will have to be taken into consideration:

- the wallpaper to be used is standard width at 540 mm wide and comes in 10 m length rolls. The pattern recurs every 500 mm so this must be accounted for in any calculations
- the cost of wallpapering is £10 per roll and the price of each roll is £10.95
- the packets of size cover 10 m^2 per packet
- the coverage of 5 L of emulsion is 35 m^2
- the coverage of 1 L of primer is 5 m^2 of bare wood
- the coverage of 2.5 L of undercoat is 10 m^2
- the coverage of 1 L of gloss coat is 6 m^2
- artex covers 1 m^2 per 1 kilogram
- cost of applying artex is £1.75 per 1 m^2
- coving wastage of 15%
- cost of fitting coving is £0.95 per 1 m
- preparation cost of wallpaper stripping and sizing of £0.45 per 1 m^2
- preparation cost of cleaning the ceiling is £0.25 per 1 m^2
- preparation cost for wood of £1.45 per 1 m^2 for stripping and priming;
- cost of applying undercoat £1.75 per 1 m^2 or £0.50 per linear metre
- cost of applying gloss coat and emulsion is £1.50 per 1 m^2 or £0.40 per linear metre
- the coverage of walls is calculated in metres squared
- the coverage of windows, skirting boards, narrow board widths is calculated in linear metres
- an overheads percentage of 8% to be added to the final total
- VAT on all materials and labour of 17.5%.

1.1 Sketching the room

Penny and David begin by drawing sketches of the room writing in the dimensions which they measured on their first visit. They draw the following elevations of the walls and plan view of the room:

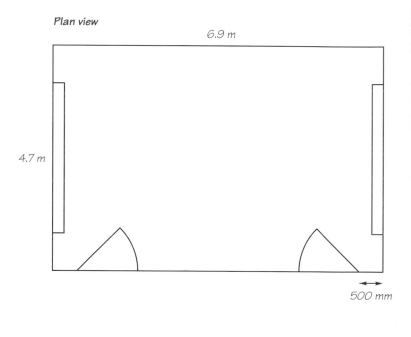

Plan view

6.9 m

4.7 m

500 mm

Elevation of partition wall

0.75 m

1.95

2.45 m

6.9 m

Elevation of end wall

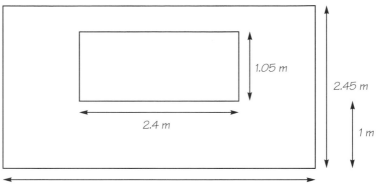

Reminder

A plan is a bird's eye view of the room and elevations show the side views of the walls. If you would like more help see Modules 5 and 28.

Practice opportunity

Make sketches of the plan and wall elevations of a room which has the following dimensions:

Length = 8.5 m
Width = 6.7 m
Height = 2.45 m

The longer walls are plain and flat, one with an entrance door 0.75 m wide by 1.95 m high in a central position. The end walls each have a window which measures 2.3 m long by 1.3 m high, also centrally placed.

1.2 Stripping and sizing the walls

Penny and David start by costing the stripping and sizing of the walls.

Calculating the surface area of the plain wall

The walls are all rectangular and so Penny and David use the formula for calculating the area of a rectangle:

$$\text{Area of a rectangle} = \text{length} \times \text{width}$$

$$\text{Area of plain wall} = \text{its length} \times \text{its width}$$

$$= 6.9 \text{ m} \times 2.45 \text{ m}$$

Reminder

When using a formula expressed in words, you substitute the quantities and do the calculation. If you would like more help with area formulas see Modules 15 and 16. For help with formulas, see Module 27.

Reminder

'Rounding' numbers or quantities gives easier amounts to calculate with. In this example the numbers have been rounded to the nearest whole number. To do this, look at the digit in the first decimal place. If it is over five, round up; if it is less than five, round down (see Module 14).

Whenever Penny and David do calculations, they always use some form of checking procedure to ensure they are correct. When they are working together, one does the calculation and the other does a checking procedure.

For this sum, Penny does an **approximate calculation** to use as a checking procedure. She rounds 6.9 up to the nearest whole number, 7, and rounds 2.45 down to 2. This gives her the following approximation:

$$7 \times 2 = 14$$

As she has rounded down more than she has rounded up, Penny says that the answer should be a little more than 14. David uses his calculator and enters the quantities using the following keying sequence:

Reminder

Numbers are entered into calculators in the order that they are read across the page. Decimal points are entered in the correct position in the number. If you would like help with using a calculator see Module 4.

The display reads **16.905**, which is sufficiently close to the approximate calculation done by Penny.

As an added check David uses another method. He divides the display number by one of the quantities which were multiplied together.

He chooses the width of 2.45 m and does the following calculation:

$$16.905 \div 2.45$$

He gets 6.9, which confirms the calculation.

The area of the plain wall is **16.905 m²**.

Reminder

Area is always measured in squared units of measurement, e.g. m², mm². If you would like further help with area see Modules 15 and 16.

Calculating the surface areas of the other walls

For the calculations of the other three walls in the room, Penny and David need to calculate the surface area of the doors and the windows.

The **approximate calculation** for a door is easy as 1.95 rounds up to 2. So 2×750 mm $= 1500$ mm; 1500 mm is the same as 1.5 m.

Using the calculator, Penny changes 750 mm into 0.75 m so that the units of measurement are the same and gets the following:

$$\text{the area of one door} = 0.75 \text{ m} \times 1.95 \text{ m}$$

$$= 1.4625 \text{ m}^2$$

This is certainly close to the approximation. They continue their calculations using this figure. Each calculation is checked and agreed.

$$\text{Area of wall with two doors} = 16.905 \text{ m}^2 - 2 \times (1.4625 \text{ m}^2)$$

$$= 16.905 \text{ m}^2 - 2.925 \text{ m}^2$$

$$= \mathbf{13.98 \text{ m}^2}$$

Penny and David calculate the area of the window and the smaller wall in the same way. Here are their final, checked calculations:

$$\text{area of wall with window} = 4.7 \text{ m} \times 2.45 \text{ m} - (2.4 \text{ m} \times 1.05 \text{ m})$$

$$= 11.515 \text{ m}^2 - 2.52 \text{ m}^2$$

$$= \mathbf{8.995 \text{ m}^2}$$

$$\text{The total surface area of the four walls} = 16.905 \text{ m}^2$$
$$+ 13.98 \text{ m}^2$$
$$+ 8.995 \text{ m}^2$$
$$+ 8.995 \text{ m}^2$$

$$= \mathbf{48.875 \text{ m}^2}$$

Penny and David round this up to the nearest 1 m^2 making a total surface area of 49 m^2.

Reminder

When multiplying decimals on paper, do the multiplication without the decimal point. In a simple problem like the examples here, do an approximate calculation, e.g. $2 \times 1.5 = 3$ and place the decimal point after the first figure in the answer: $2925 = 2.925$ m^2.

Now the preparation cost can be calculated:

$$\text{cost of stripping and sizing} = \text{£0.45 per } 1 \text{ m}^2$$

$$= \text{£0.45} \times 49$$

$$= \text{£22.05}$$

VAT is at 17.5%, so VAT on £22.05 $= \text{£22.05} \times \dfrac{17.5}{100}$

$$= \text{£3.85875}$$

$$= \text{£3.86 to the nearest penny}$$

Preparation cost with VAT added $= \text{£22.05} + \text{£3.86}$

$$= \text{£25.91}$$

> ## Reminder
>
> 'Per cent' means per 100;
>
> $17.5\% = \dfrac{17.5}{100}$
>
> $= 0.175$

How many packets of size?

Packets of size cover 10 m^2 per packet. So to cover 49 m^2 they will need five packets.

Size costs £1.55 per packet.

$$\text{Total cost of size without VAT} = 5 \times \text{£1.55}$$

$$= \text{£7.75}$$

Cost of size with VAT $= \text{£7.75} \times 1.175$

$$= \text{£9.10625}$$

$$= \text{£9.11 (correct to the nearest 1p)}$$

> ## Reminder
>
> Another method of calculating a percentage increase of 17.5% is to multiply the quantity by 1.175. If you would like help with percentages see Module 10.

$$\textbf{Total cost of stripping and sizing} = \textbf{£22.05} + \textbf{£7.75}$$

$$= \textbf{£29.80}$$

With VAT added:

$$\textbf{Total cost of stripping and sizing} = \textbf{£25.91} + \textbf{£9.11}$$

$$= \textbf{£35.02}$$

Practice opportunity

Calculate the cost of stripping and sizing the walls of a room which has the following dimensions:

> Length = 8.5 m
> Width = 6.7 m
> Height = 2.45 m

The longer walls are plain and flat, one with an entrance door 0.75 m wide by 1.95 m high in a central position. The end walls each have a window which measures 2.3 m long by 1.3 m high. The cost of stripping and sizing is £0.45 per 1 m^2.

1.3 Wall-papering the walls

Penny and David use the drawings they have made of the walls to help with their calculations:

The plain wall

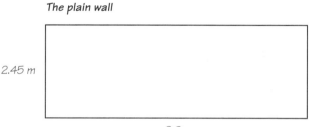

2.45 m

6.9 m

The plain wall is 6.9 m long and 2.45 m high.

The information they have about the wallpaper is that it is 540 mm wide and comes in 10 m long rolls. There is a recurring pattern every 500 mm which must be taken into consideration.

Penny and David start by finding how many widths of paper will be needed for the plain wall. They need to divide the length of the wall by the width of the wallpaper, i.e. 6.9 m ÷ 540 mm.

Penny needs to change both the dimensions into the same unit of measurement. She enters the following sum into her calculator:

> 6900 mm ÷ 540 mm

David does the approximate calculation of 7000 divided by 500; as there are two lots of 500 in each 1000, he says that

Reminder

To change metres into millimetres, multiply by 1000 as the conversion constant.

6.9 × 1000 = 6900

If you would like more help see Module 1.

the approximate answer should be 14; as the 6900 has been rounded up and the 540 has been rounded down, the answer should be just below 14.

The display of the calculator shows:

```
12.77777778
```

Penny rounds this up to 13 widths of wallpaper for the plain wall.

What effect does the pattern have?

Each drop has to measure 2.45 m. If the pattern occurs every 500 mm this means there will be five complete patterns for every drop with a small amount of wastage of 50 mm. For their calculations, it will be necessary to use a length of 2.5 m.

The total length of wallpaper required for the plain wall is 13 lots of 2.5 m = 32.5 m.

This will mean using three rolls of 10 metres and 2.5 metres from another roll.

What difference will the doors make?

The wall with the two doors will need the same amount less the area of the doors:

Each door measures 0.750 m × 1.95 m.

To calculate how many roll widths will be saved by each door, Penny does the following:

750 mm ÷ 540 mm = 1.3888888889

Penny checks this calculation by multiplying the display number of 1.3888888889 by 540 and getting the answer 750.

So, in fact, only one width of paper will be saved. The 0.388888889 will have to come out of another strip:

the length of paper saved for two doors = 2 × (1 × 1.95 m)

= 4 metres (to the nearest metre)

Considering the walls with the windows

The walls are 4.7 m by 2.45 m. How many widths of wallpaper are required to fit along the 4.7 m?

The calculation to solve this problem is 4700 mm divided by 540 mm:

$$4700 \text{ mm} \div 540 \text{ mm} = 8.703703704$$

Again Penny checks this division calculation by doing the multiplication of 8.703703704 and 540 and getting 4700.

Penny rounds this up to 9 roll widths. Each drop will be the same as for the plain walls, i.e. 2.5 m:

$$\text{total length required} = 2.5 \text{ m} \times 9$$

$$= \textbf{22.5 m}$$

Considering the windows

Each window measures 2.4 m wide × 1.05 m lengthwise. This means the window drop will save 1 m.

How many widths will fit into 2400 mm?

Penny knows that $500 \times 4 = 2000$, so she estimates that there will be just over four widths by multiplying 540 mm by 4.

With the 1 metre drop, each window will save 4×1 metre = 4 metres. two windows will save 8 metres. As the windows are 1 m from the floor and the pattern occurs every 500 mm, this will be a real saving of wallpaper.

How many rolls?

Total required for the larger walls:

$$2 \times (3 \text{ rolls} + 2.5 \text{ m}) = 6 \text{ rolls} + 5 \text{ metres}$$

Total required for smaller walls:

$$2 \times (2 \text{ rolls} + 2.5 \text{ m}) = 4 \text{ rolls} + 5 \text{ metres}$$

Total so far = 11 rolls.

From their calculations, Penny and David know that they will save 12 metres which is just over one roll. This will mean that 10 rolls of paper will be sufficient.

Reminder

0, 1, 2, 3, 4, 5, 6, 7, 8 and 9 are called 'digits'. To multiply a decimal number by 10 simply move all the digits to the next place value column on the left. if you would like more help with this see Module 12.

How many should they order?

They decide to do a quick check using the total areas of the walls and dividing it by the total area of 1 roll of paper. This is sensible because the pattern recurs at 500 mm and this fits so well with the height of the walls.

Total area of walls = 49 m^2

$$\text{area of each roll of wallpaper} = \text{width} \times \text{its length}$$

$$= 0.540 \text{ m} \times 10 \text{ m}$$

$$= 5.4 \text{ m}^2$$

$$\text{number of rolls required} = 49 \div 5.4$$

$$= 9.074$$

Because of the convenient recurring pattern in the wallpaper, they decide to order **10 rolls** which satisfies both calculations.

Calculating the cost of the wallpaper

The cost of wallpapering = £10 per roll

$$\text{for 10 rolls} = 10 \times £10$$

$$= £100$$

Reminder

'Per cent' means per 100;

$17.5\% = \dfrac{17.5}{100}$

$$= 0.175$$

With VAT this becomes £117.50.

The cost price of each roll is £10.95, so for 10 rolls the cost price is £109.50. However 17.5% VAT must be added to the cost price.

$$\text{VAT of 17.5\% on £109.50} = \frac{£109.50 \times 17.5}{100}$$

$$= £19.1625$$

$$= £19.16 \text{ to the nearest penny}$$

$$\text{total cost of wallpaper} = £109.50 + £19.16$$

$$= £128.66$$

Total cost of wallpapering = £109.50 + £100

= £209.50

With VAT added:

Total cost of wallpapering = £128.66 + £117.50

= £246.16

Practice opportunity

Calculate the cost of wallpapering the walls of a room which has the following dimensions.

Length = 8.5 m
Width = 6.7 m
Height = 2.45 m

The longer walls are plain and flat, one with an entrance door 0.75 m wide by 1.95 m high in a central position in the wall. The end walls each have a window which measures 2.3 m long and 1.3 m high.

The wallpaper costs £15.75 per roll and the charge for each roll is £12.

1.4 Cleaning and texturing the ceiling

The dimensions of the ceiling are the same as the plan view, 6.9 m by 4.7 m.

4.7 m

6.9 m

The area of the ceiling = length × width

$$= 6.9 \text{ m} \times 4.7 \text{ m}$$

$$= 32.43 \text{ m}^2$$

$$= \textbf{33 m}^2 \text{ (rounding up to the nearest whole metre)}$$

Calculating the costs of cleaning

Having calculated the area of the ceiling, Penny and David can now calculate the cost of cleaning.

The preparation cost of cleaning the ceiling is £0.25 per 1 m^2:

$$\text{preparation cost} = £0.25 \times 33$$

$$= £8.25$$

With VAT added this becomes £9.69375, i.e. £9.69 correct to the nearest penny.

Calculating the costs of the texturing material

As the ceiling is 33 m^2 it will require 33 kg to texture it; but texture is only sold in 25 kg bags and so it will require two bags, each costing £8.99.

Cost price of texturing material is 2 × £8.99 = £17.98

VAT is at 17.5%, so the selling price = £17.98 × 1.175

$$= £21.1265$$

$$= £21.13 \text{ (to the nearest 1p)}$$

Calculating the costs of application

Cost of applying texturing material is £1.75 per 1 m^2

$$= £1.75 \times 33$$

$$= £57.75$$

With VAT added this becomes £67.85625, i.e. £67.86 (to the nearest 1p).

Total cost of cleaning and texturing

$$= £8.25 + £17.98 + £57.75$$

$$= £83.98$$

With VAT added:

Total cost of cleaning and texturing

$$= £9.69 + £21.13 + £67.86$$

$$= £98.68$$

Practice opportunity

Calculate the cost of cleaning and texturing the ceiling of a room which has the dimensions:

Length $= 8.5$ m
Width $- 6.7$ m

Use the prices and information given at the beginning of this project.

1.5 Fitting the coving

Reminder
'Perimeter' is the name given to the distance all the way around the room. The perimeter is calculated by adding the lengths of the walls. If you would like more help with calculating perimeters see Modules 2 and 7.

To calculate the amount of coving, David and Penny need to calculate the perimeter of the room.

$$\text{Perimeter of room} = 2 \times 6.9 \text{ m} + 2 \times 4.7 \text{ m}$$

$$= 13.8 \text{ m} + 9.4 \text{ m}$$

$$= 23.2 \text{ m}$$

Due to the cutting process a 15% wastage factor is included for coving. They needed to calculate 115% of 23.2 m:

$$115\% \text{ of } 23.2 \text{ m} = 26.68 \text{ m}$$

The coving is supplied in 10 m packs and so they will require three packs.

Calculating the cost of the coving

One pack of coving cost £15.95:

$$\text{three packs cost} = 3 \times £15.95$$

$$= £47.85$$

VAT is charged at 17.5%.

$$\text{With VAT, three packs cost} = £47.85 \times 1.175$$

$$= £56.22375$$

$$= £56.22 \text{ (to the nearest 1p)}$$

The cost of fitting the coving is £0.95 per metre length

$$\text{cost of fitting} = 23.2 \times £0.95$$

$$= £22.04$$

$$\text{With VAT the cost of fitting} = £22.04 \times 1.175$$

$$= £25.897$$

$$= £25.90 \text{ (correct to the nearest 1p)}$$

Total cost of fitting the coving = £47.85 + £22.04

$$= £69.89$$

With Vat added:

Total cost of fitting coving = £56.22 + £25.90

$$= £82.12$$

Practice opportunity

Calculate the cost of fitting coving to a ceiling which has the dimensions of 8.5 m by 6.7 m. Use the information given for cost and fitting at the beginning of the project.

1.6 Painting the ceiling and the coving with emulsion

The cost of applying emulsion is £1.50 per 1 m^2 or £0.40 per linear metre. The coving is costed per linear metre and the ceiling is costed per squared metre.

The area of the ceiling = 32.43 m^2

The length of coving = 23.2 m.

The job specification is for the ceiling and the coving to be given three coats of emulsion.

David does an estimation to check Penny's work on the calculator. He does $3 \times 32 = 96$ and $3 \times 23 = 69$.

The total surface area to be covered = 3×32.43 m^2

$$= \mathbf{97.29 \ m^2} \ \textit{It checks!}$$

The total length to be covered = 23.2 m $\times$ 3

$$= \mathbf{69.6 \ m} \qquad \textit{It also checks!}$$

How much emulsion?

The coverage of 5 L of emulsion is 35 m^2. To calculate how much emulsion is needed, they decide to calculate the approximate surface area of the coving, bearing in mind that the coving has a curved profile.

The area of coving = its width $\times$ linear length

$$= 0.150 \text{ m} \times 69.6 \text{ m}$$

$$= \mathbf{10.44 \ m^2}$$

Total area to be covered = 97.29 m^2 + 10.44 m^2

$$= \mathbf{107.73 \ m^2}$$

To calculate the number of 5 L cans needed, the total area to be covered must be divided by the coverage of a 5 L can, i.e. 107.73 must be divided by 35. Using the calculator, Penny gets 3.078, which is just over three cans.

The coving has a profile like the following diagram:

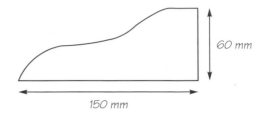

Penny calculates the area of the ceiling again taking into account the 60 mm coving width.

> ### Reminder
>
> There are 1000 mm in 1 m. To convert millimetres into metres, divide the amount by 1000. If you would like more help with converting metric measurements see Module 1.

Area of ceiling covered less the coving

$$= 6.9 \text{ m} - 0.120 \text{ m} \times 4.7 \text{ m} - 0.120 \text{ m}$$

$$= 6.78 \text{ m} \times 4.58 \text{ m}$$

$$= 31.0524 \text{ m}^2$$

Three coatings $= 3 \times 31.0524 \text{ m}^2$

$$= 93.1572 \text{ m}^2$$

This gives a new total surface area of 103.5972 m^2 (or **104 m^2**)

When this is divided by 35, the number of cans of emulsion becomes 2.95992.

Three cans should do a good job, especially as the first coat of emulsion is watered down.

Calculating the cost of applying the emulsion

Penny and David decide to use the surface area of 104 m^2 for costing the painting of the ceiling as this will be done before the wallpapering and so the coving will not cause too many problems along its edge.

Cost $= 104 \times$ £1.50

$$= £156$$

With VAT added this becomes £183.30.

Three cans of emulsion $=$ £25 $\times$ 3

$$= £75$$

With VAT this becomes £88.125, i.e. £88.13.

Total cost of painting the ceiling = £75 + £156

= £231

With VAT added:

Total cost of painting the ceiling = £88.13 + £183.30

= £271.43

Practice opportunity

Calculate the cost of painting a ceiling and coving with three coats of emulsion for a room which has the following dimensions:

Length = 8.5 m
Width = 6.7 m
Height = 2.45 m

Use the information regarding costs given at the beginning of this project and use any profile of coving which you prefer.

1.7 Working on the doors, windows, skirting boards and architrave

Reminder

It is useful to be able to do simple sums involving fractions.

$0.75 \text{ m} = \dfrac{750 \text{ mm}}{1000 \text{ mm}}$

$= \dfrac{3}{4}$

$2 \times \dfrac{3}{4} = \dfrac{6}{4} = 1\dfrac{1}{2}$

Module 22 deals with working with fractions.

All woodwork in the room is to be stripped, primed, undercoated and glossed on one side. Penny and David agree that their first task is to calculate the surface area of wood which is to be stripped and primed.

Calculating the areas

The area of one door = 0.75 m × 1.95 m

$= 1.4625 \text{ m}^2$

As a check, Penny does a quick calculation in her head. 0.75 m is $^3/_4$ m. If she rounds the height of the door to 2 m the sum becomes $2 \times {}^3/_4$, which is $1^1/_2$. This checks with the calculation.

Area of two doors = 1.4625 m^2 × 2 = 2.925 m^2

Reminder

When calculating units of measurement it is important to make sure that the same unit of measurement is being used throughout the calculation: 50 mm = 0.05 m. If you would like help converting units of measurement see Module 11.

The area of the skirting board = the perimeter of the room × the depth of the board

$$= 23.2 \text{ m} \times 0.125 \text{ m}$$

$$= 2.9 \text{ m}^2$$

The area of the skirting board = 2.9 m^2

The area of the architrave is not so simple to calculate, so David draws a diagram to help. The area of the architrave is best calculated by finding the area of the door plus the architrave and taking away the area of the door.

With the 50 mm of architrave added to the length, the total length becomes 1.95 m + 0.050 m, i.e. 2 m. With two widths of architrave around the door, the total width becomes 0.750 m + 0.050 m + 0.050 m, i.e. 0.850 m

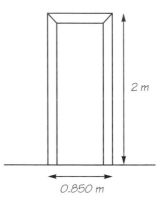

So area of door + architrave = 2 m × 0.850 m

$$= 1.7 \text{ m}^2$$

$$\text{area of architrave} = 1.7 \text{ m}^2 - 1.4625 \text{ m}^2$$

$$= 0.2375 \text{ m}^2$$

$$\text{two sets of architrave} = 2 \times 0.2375 \text{ m}^2$$

$$= 0.475 \text{ m}^2$$

Area of two sets of architrave = 0.475 m^2

Penny and David calculate the cover fillet around the window in the same way.

$$\text{area of window} = 1.05 \text{ m} \times 2.4 \text{ m}$$

$$= 2.52 \text{ m}^2$$

This time when Penny does the check she leaves the numbers as decimals and rounds the 1.05 m to 1 m and so calculates the sum $1 \times 2.4 = 2.4$, and of course it checks.

Here is the diagram of the window which they use with their calculations for the dimensions written in:

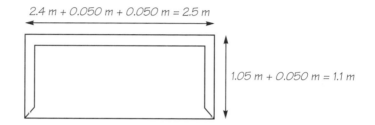

2.4 m + 0.050 m + 0.050 m = 2.5 m

1.05 m + 0.050 m = 1.1 m

$$\text{area of window plus cover fillet} = 2.5 \text{ m} \times 1.1 \text{ m}$$

$$= 2.75 \text{ m}^2$$

$$\text{area of cover fillet} = 2.75 \text{ m}^2 - 2.52 \text{ m}^2$$

$$- 0.23 \text{ m}^2$$

$$\text{area of two sets of cover fillet} = 0.23 \text{ m}^2 \times 2$$

$$= 0.46 \text{ m}^2$$

Calculating the area of wood used in the windows

Again Penny and David use a diagram to clarify their thoughts.

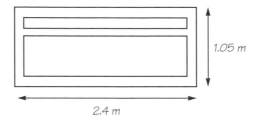

For the purpose of calculating the area of woodwork to be covered for the windows, Penny and David assume that there are three pieces of wood 2.4 m long and 50 mm wide and there are two pieces of wood 1.05 m long and 50 mm wide.

$$\text{Area of wood in one window} = 3(2.4 \text{ m} \times 0.050 \text{ m})$$
$$+ 2(1.05 \text{ m} \times 0.050 \text{ m})$$

$$= 3 \times 0.12 \text{ m}^2 + 2 \times 0.0525 \text{ m}^2$$

$$= 0.36 \text{ m}^2 + 0.105 \text{ m}^2$$

$$= 0.465 \text{ m}^2$$

$$\text{Area of wood in 2 windows} = 2 \times 0.465 \text{ m}^2$$

$$= \mathbf{0.93 \ m^2}$$

Total surface area to be covered

$$= \text{area of doors}$$
$$+ \text{area of architrave}$$
$$+ \text{area of cover fillet}$$
$$+ \text{area of wood in window}$$
$$+ \text{area of skirting board}$$

$$= 2.925 \text{ m}^2 + 0.475 \text{ m}^2 + 0.46 \text{ m}^2 + 0.93 \text{ m}^2$$
$$+ 2.9 \text{ m}^2$$

$$= \mathbf{7.69 \ m^2}$$

How many tins?

The coverage of 1 L of primer is 5 m^2 of bare wood. Two tins will be needed.

The coverage of 2.5 L of undercoat is 10 m^2, so one tin will be sufficient.

The coverage of 1 L of gloss is 6 m^2, so two tins will be needed.

Calculating the cost of materials

Two tins of primer @ £3.59 per tin = £7.18

One tin of undercoat @ £7.79 = £7.79

Two tins of gloss @ £4.45 = £8.90

Total cost = £23.87

With VAT added this becomes £28.04725, i.e. £28.05 correct to two decimal places.

They will leave any paint that is left over with the customer.

Calculating the cost of stripping and priming

The preparation cost of stripping and priming wood is £1.45 per 1 m^2

Cost of stripping and priming wood = £1.45 × 7.69

= £11.1505

= £11.15 correct to two decimal places

With VAT added this becomes £13.10125, i.e. £13.10

Calculating the cost of applying paint

The cost of applying the undercoat and the gloss is calculated in two different ways. The cost of applying to doors is calculated by the surface area, i.e. per 1 m^2.

The cost of applying to narrow strips of wood is costed by the linear metre due to the care which has to be taken along the edges.

The area of the doors $= 2.925 \text{ m}^2$

$$\text{Cost of applying to doors} = £1.75 \times 2.925 + £1.50 \times 2.925$$

$$= £5.11875 + £4.3875$$

$$= £9.50625$$

$$= £9.51 \text{ to the nearest penny}$$

With VAT added this becomes £11.17425, i.e. £11.17

$$\text{Total linear length} = \text{length of skirting board}$$
$$+ \text{ length of architrave}$$
$$+ \text{ length of coverfillet}$$
$$+ \text{ length of wood in windows}$$

$$= 23.2 \text{ m} + 2(2 \times 2 \text{ m} + 0.850 \text{ m})$$
$$+ 2(2 \times 1.1 \text{ m} + 2.5 \text{ m})$$
$$+ (2 \times 1.05 \text{ m}) + (3 \times 2.4 \text{ m})$$

$$= 23.2 \text{ m} + 2(4.850 \text{ m}) + 2(4.7 \text{ m})$$
$$+ 2.1 \text{ m} + 7.2 \text{ m}$$

$$= 23.2 \text{ m} + 9.7 \text{ m} + 9.4 \text{ m} + 2.1 \text{ m}$$
$$+ 7.2 \text{ m}$$

$$= 51.6 \text{ m}$$

$$\text{Cost of applying to wood} = 51.6 \times £0.50 + 51.6 \times £0.40$$

$$= £25.8 + £20.64$$

$$= £46.44$$

With VAT added this becomes £54.567, i.e. £54.57

Total cost of working on the wood :

Cost of applying to boards = £46.44

Cost of applying to doors = £9.51

Cost of preparation of wood = £11.15

Cost of materials = £23.87

Total cost of working on the doors, windows, skirting boards and architrave = £90.97

With VAT added:

Cost of applying to boards = £54.57

Cost of applying to doors = £11.17

Cost of preparation of wood = £13.10

Cost of materials = £28.05

Total cost of working on the doors, windows, skirting boards and architrave = £106.89

Practice opportunity

Calculate the cost of working on the doors, windows, skirting boards and architrave of a room whose dimensions are:

Length = 8.5 m

Width = 6.7 m

Height = 2.45 m

The longer wall are plain and flat, one with an entrance door 0.75 m wide and 1.95 m high in a central position. The smaller end walls each have a window which measures
2.3 m long by 1.3 m high. The architrave and the coverfillet measures 50 mm and the skirting board is 150 mm.

Assume that the prices and costings of the job are the same as for this project.

1.8 Preparing the estimate

Penny and David have to carry out two sets of calculations. One estimate is without the VAT added and the second will be with VAT. For each of these, they set out their calculations in three columns. The first column is the cost of the materials. The second column is the labour charge for each job. In the third column they put their sub-total calculated for each job. In this way, they add up the totals for the materials and for the labour and this should agree with the total of all the sub-totals of the estimate.

Job	Materials	Labour	Total
Stripping and sizing the ceiling	7.75	22.05	29.80
Wallpapering	100.00	109.50	209.50
Cleaning and texturing	17.98	66.00	83.98
Fitting coving	47.85	22.04	69.89
Painting ceiling	75.00	156.00	231.00
Woodwork	23.87	67.10	90.97
Totals	272.45	442.69	

Penny adds 272.45 and 442.69 and gets 715.14
Meanwhile David adds the **total** column and also gets 715.14
Their totals agree. No mistakes so far.
They then set about doing all the costs with VAT added:

Job	Materials	Labour	Total
Stripping and sizing the ceiling	9.11	25.91	35.02
Wallpapering	117.50	128.66	246.16
Cleaning and texturing	21.13	77.55	98.68
Fitting coving	56.22	25.90	82.12
Painting ceiling	88.13	183.30	271.43
Woodwork	28.05	78.84	106.89
Totals	320.14	520.16	

This time David adds 320.14 and 520.16 and gets 840.30
Meanwhile Penny adds the **total** column and gets 840.30 as well. Again no mistakes.

There is a final check which they do and that is calculate VAT on the total of their first estimate.

They get the figure 840.2895 which they were pleased with because many of the individual calculations have either been rounded up or rounded down to the nearest penny.

They now prepare an invoice for their customer:

MartinCrafts: Estimate for Mr & Mrs Humphreys

Job	Materials	Labour	Total
Stripping and sizing the ceiling	7.75	22.05	29.80
Wallpapering	100.00	109.50	209.50
Cleaning and texturing	17.98	66.00	83.98
Fitting coving	47.85	22.04	69.89
Painting ceiling	75.00	156.00	231.00
Woodwork	23.87	67.10	90.97
Totals	272.45	442.69	715.14
With VAT of	47.68	77.47	125.15
	320.13	520.16	840.29

There is one final calculation which needs to be done before the estimate is complete and that is the addition of an 8% overheads charge.

Plus	8% overheads charge	907.51

Their work is complete and the invoice is presented to their manager for checking and allocation of time to carry out the job.

Practice opportunity

Prepare an estimate for your room for which you have done all the calculations. Present the totals for materials and labour separately and do two sets, one with VAT added and one without.

project 2 Designing a staircase

A Number skills

Addition	Modules 2, 4, 7
Algebra	Module 27
Approximating and estimating	Modules 14
Checking procedures	Modules 12, 13, 14, 15
Division	Module 13
Fractions	Module 22
Money	Module 12, 13
Multiplication	Modules 9, 12
Percentages	Modules 4, 10
Pythagoras	Module 3
Scales and scale drawing	Modules 25, 26
Subtraction	Modules 2, 7
Technical drawing	Modules 5, 26
Trigonometry	Module 6
Using a calculator	Module 4
Using statistics	Module 20
Volumes	Modules 9, 28

B Job description and analysis of tasks

MartinCrafts have a reputation for constructing purpose built wooden components. For a local building development of five luxury houses, the builder has asked the company to estimate for constructing their five main staircases. The staircase is wider than those generally supplied by building suppliers. Julia and Keith have been given the job of designing the staircase and preparing an estimate for the builders.

Dimensions

- the height of the first floor is 2600 mm
- the maximum going is 2955 mm
- the width of the stairwell is 1100 mm
- a hand rail should be fitted to the right-hand side wall, running the full length of the stairwell.

The preliminary visit done by Julia and Keith confirmed the dimensions and they estimated the length of the strings to be approximately 3940 mm. They drew up the following list of tasks which they would have to carry out to prepare the estimate:

- calculate the rise and going of each step
- confirm the angle of the staircase to ensure that it complies with the Building Regulations
- calculate the actual length of the string using Pythagoras
- do a scale drawing of the stairwell, using side elevation and plan views
- prepare a cutting list for the five staircases
- locate suitable woods and calculate their costs
- prepare the estimate for the five flights.

C Cost of materials

English oak   £1849.00 per 1 cubic metre
Douglas fir   £1100 to £1275 per 1 cubic metre
Pirana pine   £1200 per 1 cubic metre

D Other considerations

- the builder would like to use wood which has been produced in a sustainable way by an environmentally aware company
- tread is the name given to the horizontal member of a step
- riser is the name given to the vertical member of a step
- the going is the horizontal distance between two risers
- the string is the board on either side of a stair into which the treads and the risers are fixed or housed

- it is essential that the rise and going of a stair are constant throughout the flight as this is vital for the safety of the intended users
- the Building Regulations state the following conditions for staircases:

Definition of stair	Maximum rise in mm	Minimum going in mm
Private use	220	220
Common use	190	240
Assembly areas and institutional buildings	180	280

- the Building Regulations state that twice the distance of each riser added to the going must lie between 550 mm and 700 mm
- the maximum pitch allowed for private use is 42°
- private stairs with access to more than one room must be 800 mm or over
- if a flight of stairs runs for over 1 metre, it is recommended that there should be a handrail on both sides
- the recommended height for fixing a handrail is 840 mm to 1000 mm
- 10% to be added to the cubic metreage for wastage
- 17.5% to be added for VAT.

2.1 Calculating the riser and the going

The first task when designing a staircase is to calculate the individual riser and going for each step in relation to the space available.

Reminder

'Horizontal' means parallel to the horizon. The horizontal is at right angles, or perpendicular to the vertical. 'Vertical' is always perpendicular to the horizontal.

Riser (vertical member of step)

Going (horizontal member of step)

Julia and Keith found there were two main important considerations written in the Building Regulations:

(i) the maximum rise permitted for each step in any private staircase is 220 mm;
(ii) the individual steps in a staircase have to satisfy the relationship 'twice the distance of the rise of each step added to the going must lie between 550 mm and 700 mm', i.e. 700 mm ≥ 2*R* + *G* ≥ 550 mm, where *R* is the depth of the riser and *G* is the length of the going.

Julia and Keith have to think carefully about the second condition. They rewrite the words using their own symbols, *R* for the 'riser' and *G* for the 'going' measurements and without using ≥ , 'greater than or equal to':

$$2R + G = 550 \text{ mm} - 700 \text{ mm}$$

The lower limit of the sum is 550 mm and its upper limit is 700 mm. If the number is outside this range the stairs do not satisfy the Building Regulations.

Calculating the minimum number of risers

In the house the total rise available is 2600 mm and the total going is 2995 mm.

Total rise = 2600 mm

Total going = 2995 mm 2955mm TEXT BOOK ERROR SEE PAGE 32.

The minimum number of risers to cover the height is calculated by dividing the total rise by the maximum rise permitted. Julia writes this in words:

Minimum number of risers = total rise ÷ maximum rise permitted

The measurements are substituted for the words, making sure that the same unit of measurement is used for both amounts:

$$= 2600 \text{ mm} \div 220 \text{ mm}$$

Julia starts by doing an **approximate calculation.** This will be used as a checking procedure for the calculation done by the calculator.

She uses her 10× table:

220 mm × 10 = 2200 mm

There are at least 10 risers, but how many more?

She then subtracts this amount from the total rise:

2600 mm − 2200 mm = 400 mm

This means there will be 11 risers, but not 12 because 2 × 220 mm = 440 mm, and there is only 400 mm left.

In the meantime, Keith enters the figures into his calculator and gets the following display:

| 11.81818182 |

This agrees with Julia's approximate calculation.

They decide to round the number up to the nearest whole number, 12.

For private houses, Julia and Keith know that 220 mm is the maximum allowable depth for a riser. Keith writes:

Actual rise if 12 risers are used = total rise divided by 12

= 2600 mm ÷ 12

This time Keith does the approximate calculation for the checking procedure. He does 12 × 200 = 2400, which leaves 200 mm. He calculates 12 × 10 = 120 and adds this to 2400 and gets 12 × 210 = 2520. The answer is going to be over 210 mm.

In the meantime, Julia uses the calculator and gets the following display:

| 216.6666667 |

This is 217 mm to the nearest millimetre.

Keith and Julia realise that this is getting very close to the permitted maximum.

They decide to try 13 risers.

Actual rise if 13 risers used = 2600 mm ÷ 13

= 200 mm

They feel that this is an appropriate size to work with.

How many treads?

The number of treads required is always one less than the number of risers because the floor is one tread.

The number of treads required for this stair case = 13 − 1

= 12

The total possible going measures 2955 mm.

Again Julia wrote in words the calculation which needed to be done:

Actual going of each step = total possible going ÷ 12

= 2955 mm ÷ 12

For a checking procedure, she did the approximate calculation doing 12 × 200 = 2400 and quickly following it with 50 × 12 = 600, making 12 × 250 = 3000. The going would be just below 250 mm.

Keith got the following on his calculator display:

246.25

They decide to assume a going of 245 mm as this is over the minimum measurement allowed.

Reminder

When you are giving answers correct to the nearest whole number, and the amount has a decimal fraction part, you look at the digit in the first decimal place. If this is greater than five you round up to the next whole number. If it is less than five you round the number down to the whole number only. (Module 14)

Reminder

There are other checking procedures which would do just as well as the approximate calculations. e.g. inverse operations: If 2955 divided by 12 gives the answer 246.25, it is possible to use the inverse operation of multiplication as a check. Multiply the answer by the number you divided by: 246.25 × 12 = 2955. *It checks!*

Reminder

If you would like more help with substitutions see Module 27.

Checking the Building Regulations

These calculations have to be checked to show that they satisfy the Building Regulations. The calculated measurements are substituted into the expression:

$$2R + G = 2 \times 200 \text{ mm} + 245 \text{ mm}$$

$$= 400 \text{ mm} + 245 \text{ mm}$$

$$= 645 \text{ mm}$$

This figure is within the acceptable limit of 550 mm to 700 mm and so they can proceed with the calculations.

Practice opportunity

Calculate the riser and going of a staircase which is to be constructed in a space which allows a total going of 3450 mm and a total rise of 3000 mm. Ensure that the necessary Building Regulations will be met by your design.

2.2 Calculating the slope of the staircase

Julia and Keith know the maximum slope of a staircase for private usage is limited to 42°. The slope of a staircase is determined by the angle of pitch. They can calculate this by using the measurements of the total rise and the total going. Their staircase diagram shows that it can be represented by a right-angled triangle:

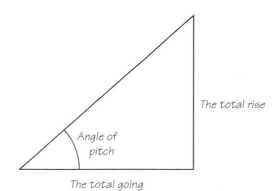

The total rise

Angle of pitch

The total going

Calculating the total rise and the total going

There are going to be 13 risers each of 200 mm. They found this very easy to calculate:

$$200 = 2 \times 100$$

so 13 must be multiplied by 2 and then by 100.

$$13 \times 2 = 26; \ 26 \times 100 = 2600$$

The total rise of the staircase $= 13 \times 200$ mm

$$= 2600 \text{ mm}$$

If there are 13 risers, there will be 12 treads. Using their calculated figure of 245 mm for the going of each step:

The total going $= 12 \times 245$ mm

Julia remembers her earlier calculation of $12 \times 250 = 3000$. 245 is 5 less than 250, so the answer will be 5×12, i.e. 60 less than 3000.

$$3000 - 60 = 2940$$

Keith enters the calculation into his calculator. He gets the following display which checked.

Julia and Keith draw a diagram of the information so far:

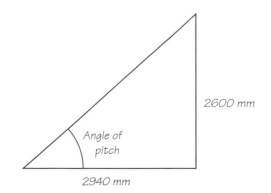

Reminder

If you would like help with using the tangent ratio see Module 6.

Reminder

$\frac{2600}{2940}$ is a fraction.

To express it as a decimal you have to divide its numerator, 2600 by its denominator of 2940. If you would like help with fractions see Module 22.

Using the tangent ratio

Julia and Keith decide it is time to use trigonometry. They know the lengths of the two shorter sides in the triangle. This means that they can use the tangent ratio to calculate the angle of pitch of the staircase.

Using trigonometry:

$$\text{Tangent of the angle of pitch} = \frac{\text{length of the side opposite}}{\text{length of the adjacent side}}$$

$$= \frac{2600}{2940}$$

Keith enters this division sum into the calculator and gets the following display:

0.884353741

Leaving this display in the calculator, Keith presses the key

$\tan^{-1}$

and gets the following:

41.48805735°

Keith and Julia check the calculation by doing everything in reverse order:

Leaving the display of 41.48805735°, Keith presses the tan key and gets his earlier display of 0.88435741. He then multiplies by the number he had divided by, i.e. 2940 and gets the display 2600. So far so good.

An angle of pitch of 41.5° is only just acceptable, so they will have to ensure that the goings are measured accurately.

Practice opportunity

Calculate the angle of the staircase which has a total going of 3450 mm and a total rise of 3000 mm. Or design a staircase for a given space and calculate its angle of pitch.

2.3 Calculating the string length

If the total rise of the staircase is going to be 2600 mm and the total going is to be 2940 mm, how can the length of the string be calculated?

Julia and Keith decide to use Pythagoras' theorem.

The length of the string is the hypotenuse of the right-angled triangle which is formed by the total rise and the total going. They draw a diagram of the problem:

Reminder

The side opposite to the right angle in any right-angled triangle is called the hypotenuse. Pythagoras' theorem is the area of the square drawn on the hypotenuse is equal to the sum of the areas which are drawn on the two shorter sides. See Module 3.

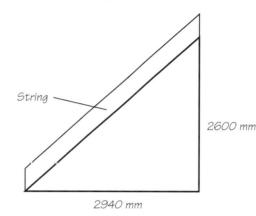

Using Pythagoras:

$$(\text{Length of string})^2 = (2600^2 + 2940^2)$$

$$= 15403600$$

Having found the area of the square on the hypotenuse, Julia and David set about finding the length of the hypotenuse and hence the length of the string:

$$\text{Length of string} = \sqrt{(2600^2 + 2940^2)}$$

$$= \sqrt{(6760000 + 8643600)}$$

$$= \sqrt{15403600}$$

$$= 3924.74203 \text{ mm}$$

$$= 3925 \text{ mm to the nearest millimetre}$$

Reminder

When the area of the square on the hypotenuse has been calculated, you can calculate the length of the side of the square by using the square root function on your calculator. See Module 4 for help with calculating square roots using a calculator.

As this is an important calculation, Julia and David do a checking procedure using their calculators. Leaving the answer on their displays of 3924.74203, they use the

	x^2

key and check that it returns to 15403600.

The length of the handrail would be the same.

Practice opportunity

Calculate the length of the string for a straight staircase which has been designed to fit into a space which allows a total going of 3450 mm and a total rise of 3000 mm.

2.4 Scale drawing of the stairs

Keith and Julia prepare to draw a scaled drawing of their design. They decide to use the scale of 1:100 to convey the general construction. They begin by making a list of the dimensions of the wood to be used in the construction.

The finished sizes of the different parts of the stairs are agreed as:

Treads	length 1100 mm	width 245 mm
Risers	length 1100 mm	width 200 mm
Strings	length 3925 mm	width 295 mm
Handrails	length 3925 mm	width 70 mm

Using a scale of 1:100, each measurement on the drawing will be

$$\frac{1}{100}$$

Reminder

Scale ratios describe the proportions of quantities or lengths. The ratio 1:100 means that on the drawing 1 mm represents 100 mm of the full size. See Module 26 for further help.

of its actual measurement. The total rise of the staircase, 2600 mm will measure 26 mm on the drawing and the total going, 2940 mm will measure 29.4 mm. It will be difficult to do an accurate measurement for any fractional parts of a millimetre.

Opposite are their drawings with each of the risers numbered in the elevation and each of the treads numbered in the plan view:

Practice opportunity

Prepare a list of the wood to be used in the construction of a flight of stairs and draw a scaled drawing of their design. Choose an appropriate scale for your drawing.

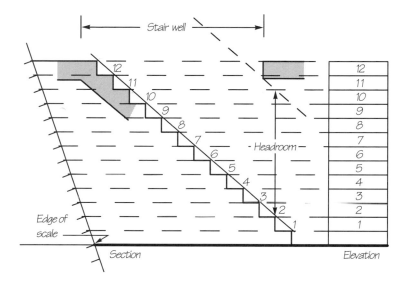

Section

Elevation

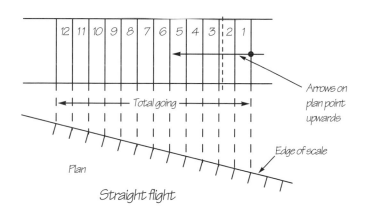

Straight flight

2.5 Preparing the cutting list

Julia and Keith start by making a list of the individual members of the staircase and how many are required of each for one complete staircase:

treads	12
risers	13
strings	2
handrails	2

They then calculate the total number of parts required for the five flights of stairs and draw up a cutting list, giving the measurements in millimetres:

Job title: Straight flight of stairs × 5 (Temple Stream)

Item	No. off	Length	Sawn size		Finished size	
			W	TH	W	TH
Strings	10	3925	300	40	295	35
Treads	60	1100	250	30	245	25
Risers	65	1100	205	20	200	15
Handrails	10	3925	75	50	70	45

Practice opportunity

Draw up a cutting list of materials required to make a flight of stairs which you have designed.

2.6 Calculating costs

There are two options which have to be followed to provide the customer a choice between soft and hard wood.

The price of the wood

Julia and Keith have a list of suppliers of softwoods who have been approved by the Association of Environmentally Approved Suppliers of Timber.

The prices they are quoted for Douglas fir are £1150, £1275, £1100. All the prices are subject to availability and fluctuation in the market. They decide to calculate an average or mean price and use it for their estimate.

$$\text{Average price} = \frac{£1150 + £1275 + £1100}{3}$$
$$= £1175$$

The average price of Douglas fir per 1 m^3 is approximately £1175.

They also price the job using Welsh oak from similar sources and get an average price of £1800 per 1 cubic metre.

Calculating the volumes

Costs are calculated from the cutting list. The sawn dimensions are used to calculate the volumes of wood as this is the amount of material actually ordered. Here are the volume calculations done by Julia and Keith:

$$\text{The volume of 1 string} = \text{length} \times \text{breadth} \times \text{depth}$$
$$= 3925 \text{ mm} \times 300 \text{ mm} \times 40 \text{ mm}$$
$$= 3.925 \text{ m} \times 0.3 \text{ m} \times 0.04 \text{ m}$$
$$= 0.0471 \text{ m}^3$$

The checking procedure used for all these calculations is to do the sum in a different order. The check for their first calculation is to multiply the following:

$$0.3 \times 0.04 \times 3.925 = 0.0471 \quad \textit{It checks!}$$

For 10 strings this becomes:

$$\text{Volume} = 10 \times 0.0471 \text{ m}^3$$
$$= \mathbf{0.471 \text{ m}^3}$$

> **Reminder**
>
> The average which has been calculated is sometimes called the arithmetic mean. If you would like further help with averages see Module 20. The mean is calculated by adding all the items and dividing their total by the number in the set.

> **Reminder**
>
> Volume is always measured in cubic units. For rectangular shapes there will always be three dimensions to measure and use in the calculation. If you would like help with volumes see Modules 9 and 28.

$$\text{Volume of one tread} = \text{length} \times \text{breadth} \times \text{height}$$
$$= 1100 \text{ mm} \times 250 \text{ mm} \times 30 \text{ mm}$$
$$= 1.1 \text{ m} \times 0.25 \text{ m} \times 0.03 \text{ m}$$
$$= 0.00825 \text{ m}^3$$

For 60 treads this becomes:

$$\text{Volume} = 60 \times 0.00825 \text{ m}^3$$
$$= \mathbf{0.495 \text{ m}^3}$$

$$\text{Volume of one riser} = \text{length} \times \text{breadth} \times \text{height}$$
$$= 1100 \text{ mm} \times 205 \text{ mm} \times 20 \text{ mm}$$
$$= 1.1 \text{ m} \times 0.205 \text{ m} \times 0.02 \text{ m}$$
$$= 0.00451 \text{ m}^3$$

For 65 risers this becomes:

$$\text{Volume} = 65 \times 0.00451 \text{ m}^3$$
$$= \mathbf{0.29315 \text{ m}^3}$$

$$\text{Volume of one handrail} = \text{length} \times \text{breadth} \times \text{height}$$
$$= 3925 \text{ mm} \times 75 \text{ mm} \times 50 \text{ mm}$$
$$= 3.925 \text{ m} \times 0.075 \text{ m} \times 0.05 \text{ m}$$
$$= 0.01471875 \text{ m}^3$$

For 10 handrails this becomes:

$$\text{Volume} = 10 \times 0.01471875 \text{ m}^3$$
$$= \mathbf{0.1471875 \text{ m}^3}$$

Julia and Keith decide to calculate the total volume using the figures from the calculation and not using any approximations:

$$\text{Total volume} = 0.471 \text{ m}^3 + 0.495 \text{ m}^3 + 0.29315 \text{ m}^3$$
$$+ 0.1471875 \text{ m}^3$$
$$= \mathbf{1.4063375 \text{ m}^3}$$

At this stage it would seem appropriate to calculate the 10% wastage using this figure:

$$\text{Wood plus wastage allowance} = 1.4063375 \text{ m}^3 \times 1.01$$
$$= 1.420400875 \text{ m}^3$$

At this stage Julia and Keith decide to work to two decimal places: 1.42 m^3

$$\text{Cost of Douglas fir} = 1.42 \times £1175$$

Reminder

To calculate a percentage increase of 10% can be calculated by multiplying the quantity by 1.01. If you would like more help with calculating percentages see Module 10.

When this was done on the calculator the display showed 1668.5; as this is a money amount, Julia and Keith knew that the cost was:

$$= £1668.50$$

$$\text{Cost of Welsh oak} = 1.42 \times £1800$$

$$= £2556$$

Practice opportunity

Calculate the cost of wood for the staircase which you have designed. You should check current prices of wood and the availability of wood which has been produced using sustainable methods.

2.7 Preparing the estimate

Estimates are always checked carefully at MartinCrafts because this is the most important communication which is made to the customer. Julia and Keith work out the costs in two different ways. Firstly they calculate the percentage increase by multiplying by 1.175:

The cost of the wood for Douglas fir = £1668.50

With 17.5% VAT added this becomes £1960.4875

£1960.49 to the nearest penny

The cost of wood for Welsh oak = £2556

With 17.5% VAT added this becomes £3003.30

Secondly, they calculate the cost showing the VAT as a separate amount and then adding it to the given price. These two ways of calculating provide a useful check.

Here is the estimate which Julia and Keith produced:

MartinCrafts

Materials for five sets of stairs for Temple Stream

Using Douglas fir @ £1175	1668.50
17.5% VAT	291.99
Total	1960.49
Using Welsh oak @ £1800	2556.00
17.5% VAT	447.30
Total	3003.30

Practice opportunity

Produce an estimate for the cost of materials for the stairs you have designed.

project 3 Planning kitchens

A Number skills

Addition	Modules 2, 4, 7
Algebra	Module 27
Approximating and estimating	Module 14
Areas	Modules 15, 16
Checking procedures	Modules 12, 13, 14, 15
Converting units of measurement	Module 11
Division	Module 13
Fractions	Module 22
Money	Module 12, 13
Multiplication	Modules 9, 12
Percentages	Modules 4, 10
Scales and scale drawing	Modules 25, 26
Subtraction	Modules 2, 7
Technical drawing	Modules 5, 26
Using a calculator	Module 4

B Job description and analysis of tasks

MartinCrafts have been asked to quote for the job of planning and fitting 15 kitchens with wall and floor units in a development of town terraces. Each house is the same so all the kitchens are identical in layout. Other local companies will be supplying and fitting the cookers, washing machines and fridges and plumbing in the sink. Susmita and Joe are given the task of preparing the estimate for the customer.

Dimensions

- each kitchen floor is 2800 mm by 3900 mm
- the height of the kitchens is 2700 mm
- the builder has positioned the sink and installed power points for the washing machine, cooker and refrigerator

- the cookers are to measure 600 mm by 500 mm by 1200 mm
- the fridges are to measure 600 mm by 550 mm by 850 mm
- the washing machines are to measure 600 mm by 550 mm by 850 mm.

The preliminary visit done by Susmita and Joe confirmed the dimensions and the positions of the power points for the electrical appliances. They draw up the following list of tasks which they will have to carry out to prepare the estimate:

- draw the plan of the kitchen showing power points, water supply, doors and windows
- position the cooker, washing machine and fridge at the appropriate power points
- plan the number of units required
- make a list of the requirements of specialist fittings, work surfaces, etc.
- calculate the cost of the units
- calculate the cost of the specialist fittings, work surfaces
- do an axonometric drawing of the kitchen to present to the client
- prepare the estimate.

C Cost of materials

	Width × Depth × Height	Cost in £ and p
Single wall unit	600 mm × 300 mm × 600 mm	192.26
Double floor unit for sink	1000 mm × 500 mm × 882 mm	287.57
Double floor unit	800 mm × 500 mm × 882 mm	407.23
Broom cupboard	600 mm × 500 mm × 2092 mm	450.98
Single four-drawer unit	600 mm × 500 mm × 882 mm	356.29
Corner unit	635 mm × 635 mm × 882 mm	305.78
Frontal pack for the washing machine		258.64
Frontal cover for fridge		213.53
Cornice		17.79 per 2700 mm
Plinth		61.25 per 2700 mm
Worktops		64.25 per 2700 mm
Drawer or cupboard knobs		92p each

D Other considerations

- the distance between the cooker and the sink should ideally be no less than 1200 mm and no more than 1800 mm
- the path between the cooker and the sink should not be across a main walking route through the kitchen for safety reasons
- there should be worktop space either side of the sink
- the arrangement of the cooker and the sink should include worktop surfaces between and be something like, for example, worktop, cooker, worktop, sink, worktop
- the total distance of the perimeter of the work triangle from the fridge to the sink to the cooker should lie between 3600 mm and 6600 mm
- labour of £125 is to be estimated for the fitting of each set of planned units per kitchen
- the supplier gives a 40% trade discount for bulk orders
- a 10% wastage factor is used for materials such as coving, plinth and worktop
- a 10% profits and overheads figure is to be added to the final costing total
- 17.5% VAT is to be added to the final cost of components and labour.

3.1 Drawing a scaled plan of the kitchen

Susmita and Joe start by doing a scaled plan of the kitchen. They use the scale of 1:50.

The width of the kitchen is 2800 mm; using a scale of 1:50, every 50 mm is represented by 1 mm on the plan.

Calculating scaled distances

To calculate any distance on the scaled drawing using the scale 1:50, they will have to divide the actual measurements by 50.

$$
\begin{array}{r}
2800 \div 50 \\
6 \\
50 \\
\hline
50/2800 \\
2500 \quad (50 \times 50) \\
\hline
300 \\
\hline
300 \quad (50 \times 6)
\end{array}
$$

Reminder

Drawing scales are written as unitary ratios. The ratio 1:50 means that on the drawing 1 mm represents 50 mm of the full size. See Module 26 for further help.

Reminder

The method of division used here is called subtraction division. The numbers above the line have to be added together for the answer. If you would like more help with division see Module 13.

Reminder

It is good practice to use checking procedures for any calculations. If you would like more help with using a check see Modules 12, 13, 14 and 15.

The width of 2800 mm will measure 56 mm on the scaled plan (2800 ÷ 50).

Susmita and Joe always use a checking procedure to make sure that their calculations are correct. Susmita does a mental calculation, dividing 2800 first by 100 and then multiplying her answer by 2. The combined result should be the same as dividing by 50.

2800 divided by 100 = 28

28 × 2 = 56 *It checks with their division answer.*

Here is their drawing using the measurements taken during their preliminary visit showing the dimensions, windows, positions of the power points and water inlet:

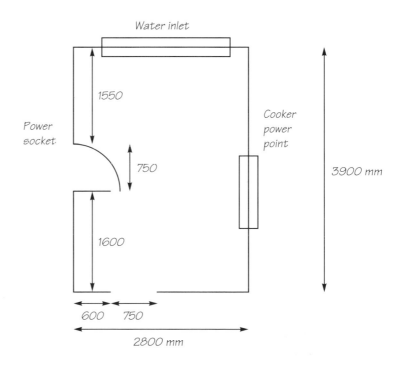

They mark in the position of the sink and then decide on suitable positions for the fridge, washing machine and cooker; here is their resulting scaled drawing. Both the fridge and the cooker are placed 900 mm from each corner:

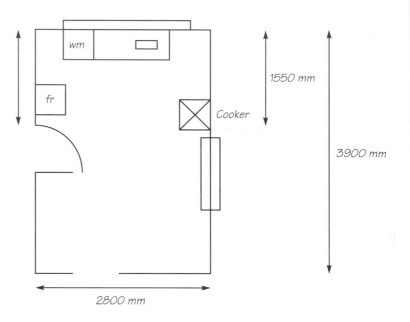

At MartinCrafts, they use the following symbols to represent appliances:

 gas or electric cooker

 fridge, refrigerator or freezer

 washing machine or wash boiler

> ## Reminder
>
> Any measurement which is taken from a scaled drawing will be approximate and not accurate. It is generally agreed that written dimensions are taken from scaled drawings for work which requires accuracy.

Measuring the work triangle

On the scaled drawing, they take approximate measurements for the work triangle between the cooker, fridge and sink. The measurements are given in metres on their scale rule:

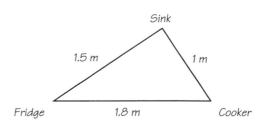

Reminder

When converting metres to millimetres, the conversion factor is 1000 because 1000 mm = 1 m. When multiplying by 1000, all the digits in the number move three place values to the left. See Module 11 for further help.

To change the metres into millimetres Susmita multiplies each measurement by a conversion factor of 1000.

$1.8 \text{ m} \times 1000 = 1500 \text{ mm}$

$1.5 \text{ m} \times 1000 = 1500 \text{ mm}$

$1.5 \text{ m} \times 1000 = 1500 \text{ mm}$

$1 \text{ m} \times 1000 = 1000 \text{ mm}$

The total distance around the perimeter of the triangle after converting the metre measurements into millimetres is:

$1800 \text{ mm} + 1500 \text{ mm} + 1000 \text{ mm} = 4300 \text{ mm}$

This total lies between the suggested limits of a good design i.e. between 3600 mm and 6600 mm.

Practice opportunity

Draw the plan of a kitchen which you have measured and calculate the distance of the work triangle between the cooker, fridge and sink.

Practice opportunity

Design a layout of a kitchen which needs to be fitted with units and appliances. Calculate the work triangle between the cooker, fridge and sink.

3.2 Planning the units

From experience, Joe and Susmita know that they can get wall units which are either 300 mm, 500 mm or 600 mm in width. Their depth will usually be 300 mm.

Here is an oblique sketch of any unit, showing what is called their width, depth and height:

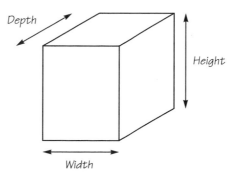

The floor units are also available in a range of widths from 300 mm to 1000 mm and are usually about 900 mm in height and 500 mm deep. There are a range of tall cupboards for special uses such as broom storage or as a larder.

Joe draws the shorter wall elevation which has a window and where the sink has been placed. The most important fact for him to remember is that the wall meets other walls at each corner and if these also have units on, then space must be given to them on his scale drawing.

Again he uses a scale of 1:50 and produces the following arrangement which is pleasing and should be possible:

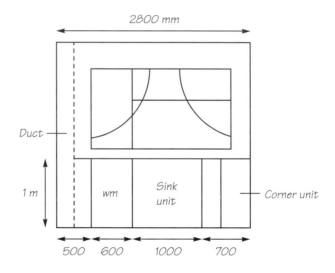

Joe checks that the widths of the units chosen will fit into the overall distance of 2800 mm.

The measurement 500 mm represents the space taken by the duct and the units on the other wall; 600 mm is the width of the washing machine. Sink units are usually 1 metre long and a corner unit of about 700 mm would complete the wall.

This enables him to make a list of kitchen units required for this wall:

1 frontal washing machine pack – 600 mm

1 double floor unit for the sink – 1000 mm

1 corner unit which fits into 700 mm

He checks that the dimensions he has chosen will fit into the width of the kitchen:

Reminder

When adding units of measurement it is important to make sure that the same unit of measurement is being used for each dimension.

$500 \text{ mm} + 600 \text{ mm} + 1000 \text{ mm} + 700 \text{ mm} = 2800 \text{ mm}$
It checks!

Susmita draws the wall elevation for the longer internal wall which has a door. She also uses a scale of 1:50:

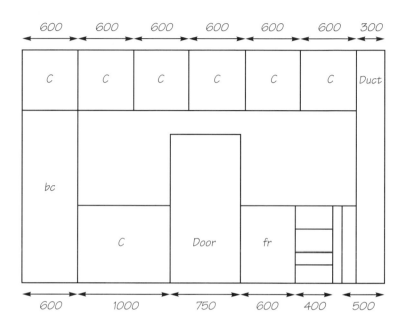

Susmita checks that the units will fit into the length of the kitchen.

The top units are all 600 mm wide. There are six of them.

$6 \times 600 \text{ mm} = 3600 \text{ mm}$

Adding this to the duct:

$3600 \text{ mm} + 300 \text{ mm} = 3900 \text{ mm}$ *It checks!*

Now for the floor units:

$600 \text{ mm} + 1000 \text{ mm} + 750 \text{ mm} + 600 \text{ mm} + 400 \text{ mm}$
$+ 500 \text{ mm} = 3850 \text{ mm}$ *It also fits with 50 mm to spare!*

Susmita's list of units required for this kitchen wall at the planning stage is as follows:

6 single wall units – 600 mm wide
1 broom cupboard – 600 mm wide

1 double floor unit, width 1000 mm
1 single floor drawer unit 400 mm wide
1 fridge front – 600 mm wide

Susmita has to make a note to check that the broom cupboard and the wall unit above it will fit.

Joe and Susmita check through their supplier's catalogue and find the following suitable units:

	Width × Depth × Height
Single wall unit	600 mm × 300 mm × 600 mm
Double floor unit for sink	1000 mm × 500 mm × 882 mm
Double floor unit	800 mm × 500 mm × 882 mm
Broom cupboard	600 mm × 500 mm × 2092 mm
Single four-drawer unit	400 mm × 500 mm × 882 mm
Corner unit	635 mm × 635 mm × 882 mm
Frontal pack for the washing machine	600 mm wide
Frontal cover for fridge	600 mm wide

Susmita sees that the height of the broom cupboard is 2092 mm and the height of the wall unit above it is 600 mm:

$$2092 \text{ mm} + 600 \text{ mm} = 2692 \text{ mm}$$

This will be a tight fit as the wall height is 2700 mm but it should be possible.

Practice opportunity

Design a kitchen layout and make a list of requirements of the units. Support your work with plan and wall elevations.

3.3 Specialist fittings

How much coving?

Coving is the name given to the finishing wood which is fitted at the top of the wall units and meets the ceiling. The wall units run along the length of the longer wall only.

Here is Jo's sketch of the top of the wall units along the wall.

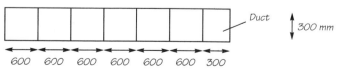

All the units have a depth of 300 mm which matches the depth of the duct in the corner.

The coving would look better if it runs the full length of the wall, i.e. 3900 mm.

How much plinth?

The plinth runs along the bottom of the floor units and the floor. To calculate the amount required, Susmita and Joe decide to draw a plan of the kitchen with all the floor units drawn in. They use a scale of 1:50:

Reminder

On the plan, the length of the bottom of the corner unit has been calculated and written in. It is always good practice to keep a record of any calculations for later reference.

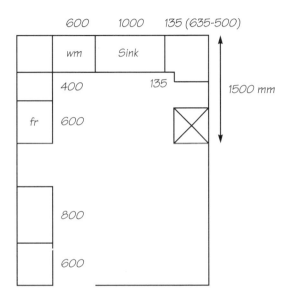

Using the plan:

Bottom left-hand corner: 600 mm + 800 mm + 500 mm (depth to wall)

$$600 + 800 + 500 = 1900$$

Reminder

Doing any calculation in a different order is a useful checking procedure. See Modules 12, 13, 14 and 15 for further help.

While Joe is doing the calculation, Susmita does a checking procedure by adding the amounts in a different order:

$$500 + 800 + 600 = 1900 \ \textit{It checks!}$$

From the door, around the sink, to the cooker: 500 mm (depth to wall) + 600 mm + 400 mm + 600 mm + 1000 mm + 135 mm + 135 mm + 500 mm (depth to wall)

$$500 + 600 + 400 + 600 + 1000 + 135 + 135 + 500 = 3870$$

Again this sum is checked by doing the additions in a different order.

$$\text{Length of plinth required} = 1900 \text{ mm} + 3870 \text{ mm}$$

$$= 5770 \text{ mm}$$

Work surfaces

Worktops are supplied in 2.7 metre lengths and are 600 mm wide and 40 mm thick.
  Using their kitchen plan:

Bottom left-hand corner:

$$600 \text{ mm} + 800 \text{ mm} = 1400 \text{ mm}$$

To the duct on the longer wall:

$$600 \text{ mm} + 400 \text{ mm} + 200 \text{ mm} = 1200 \text{ mm}$$

Along the shorter wall:

$$200 \text{ mm} + 600 \text{ mm} + 1000 \text{ mm} + 635 \text{ mm} = 2435 \text{ mm}$$

To the cooker from the corner:

$$900 \text{ mm} \ (1500 - \text{width of cooker, i.e. } 600)$$

$$\text{Total length} = 1400 \text{ mm} + 1200 \text{ mm} + 2435 \text{ mm} + 900 \text{ mm}$$

$$= 5935 \text{ mm}$$

Cupboard and drawer knobs

14 required on the long wall

5 on the short wall

$$14 + 5 = 19$$

19 required altogether

Practice opportunity

For a kitchen which you have designed calculate the requirements of specialist fittings which should include – coving, plinth, worktop and cupboard and drawer knobs.

3.4   Calculating the cost of the units

Here is the list of units and their prices required for one kitchen prepared by Susmita and Jo:

	Width × Depth × Height	Cost in £ and p
6 Single wall unit @ 192.26	600 mm × 300 mm × 600 mm	1153.56
Double floor unit for sink	1000 mm × 500 mm × 882 mm	287.57
Double floor unit	800 mm × 500 mm × 882 mm	407.23
Broom cupboard	600 mm × 500 mm × 2092 mm	450.98
Single four drawer unit	600 mm × 500 mm × 882 mm	356.29
Corner unit	635 mm × 635 mm × 882 mm	305.78
Frontal pack for the washing machine		258.64
Frontal cover for fridge		213.53
Total		**3433.58**

Susmita and Joe do the calculation individually and check that they get the same answer.

Susmita then uses her calculator to find the cost for 15 sets of units:

For 15 kitchens: cost of units $= 15 \times £3433.58$

$$= £51\ 503.70$$

Joe uses estimation for a checking procedure:

$10 \times 3500 = 35\ 000$

Now he calculates 5×3500; 5 is half of 10 so:

$5 \times 3500 = {}^{1}\!/_{2} \times 35\ 000$

$$= 17\ 500$$

The total should be approximately 52 500 (35 000 + 17 500), but just below it because he rounded up to 3500.
It checks!

The supplier will give a 40% discount on this bulk order. This makes the cost 40 % less.

Susmita calculates the 40% discount by multiplying the cost by 0.6:

Cost of units with 40% discount = £51 503.70 × 0.6

$$= £30\ 902.22$$

Joe checks this by calculating 40% of £51 503.70:

40% × £51 503.70 = £20 601.48

He then subtracts this from the total cost:

£51 503.70 − £20 601.48 = £30 902.22 *which checks!*

Practice opportunity:

Calculate the cost of the units required for the kitchen which you have designed. Remember to calculate any discount given for bulk orders or for any special offers which are around.

3.5 Calculating the cost of the specialist fittings

The coving, plinth and worktop material is supplied in lengths of 2700 mm.

Specialist materials for one kitchen

Requirements for one kitchen are:

Coving	3900 mm
Plinth	5770 mm
Work surface	5935 mm

Two lengths of coving = 2700 mm × 2 = 5400 mm

Susmita calculates how much is left over:

5400 mm − 3900 mm = 1500 mm

This gives 1500 mm of coving extra. Susmita is concerned about how much is left over and calculates it as a percentage:

Reminder

To change a fraction into a percentage, first calculate it as a decimal fraction and then multiply by 100.

$$\frac{\text{Amount left over}}{\text{Amount ordered}} \times 100\% = \frac{1500}{5400} \times 100\%$$

$$= 27.777777778\%$$

$$= 27.78\% \text{ (correct to two decimal places)}$$

$$\text{Three lengths of plinth} = 3 \times 2700 \text{ mm}$$

$$= 8100 \text{ mm}$$

Susmita calculates how much is left over:

$$8100 \text{ mm} - 5770 \text{ mm} = 2330 \text{ mm}$$

This gives 2330 mm of plinth extra. The wastage as a percentage is:

$$\frac{\text{Amount left over}}{\text{Amount ordered}} \times 100\% = \frac{2330}{8100} \times 100\%$$

$$= 28.77\%$$

$$\text{Three lengths of worktop} = 3 \times 2700 \text{ mm}$$

$$= 8100 \text{ mm}$$

Susmita calculates how much is left over:

$$8100 \text{ mm} - 5935 \text{ mm} = 2165 \text{ mm}$$

This gives 2165 mm of worktop extra. From the previous calculation for the plinth Susmita can see that the wastage factor is again very high.

Each requirement is greater than is needed and will provide sufficient extra should any problems occur, but too much is wasted for each kitchen.

Joe decides to work out the cost for each kitchen and compare the amount to calculating the material requirements for all 15 kitchens and using the wastage factor of 10% which is usual for these specialist materials.

Reminder

To write 27.77777778 correct to two decimal places, look at the digit in the third decimal place. If it is greater than five, then raise the second decimal number by one. If it is less than five then write the second decimal number as it is. See Module 14.

Reminder

When multiplying decimals, do the multiplication without any decimal points. Count the total number of digits after the decimal points in the numbers and then count in this number of places from the right of the answer. See Module 12 for further help.

Cost of materials for one kitchen

Susmita does the calculations using her calculator and Joe does a checking procedure by doing the calculation in the reverse order.

Cost of coving $= 2 \times £17.79$

$$= £35.58$$

Checking procedure: $17.79 \times 2 = 35.58$ *It checks!*

Cost of plinth $= 3 \times £61.25$

$$= £183.75$$

Checking procedure: $61.25 \times 3 = 183.75$ *It checks!*

Cost of worktop $= 3 \times £64.25$

$$= £192.75$$

Checking procedure: $64.25 \times 3 = 192.75$ *It checks!*

Cost of handles $= 19 \times £0.92$

$$= £17.48$$

Checking procedure: $0.92 \times 19 = 17.48$

Total cost for specialist fittings for one kitchen:

$$= £35.58 + £183.75 + £192.75 + £17.48$$

$$= £429.56$$

Checking procedure: $17.48 + 192.75 + 183.75 + 35.58$

$$= 429.66$$

Cost for 15 kitchens

Using the calculated cost for one kitchen:

For 15 kitchens $= 15 \times £429.56 = £6443.40$

Checking procedure: $429.56 \times 15 = 6443.4$

Calculating materials required for 15 kitchens

Joe uses his calculator and Susmita does the checks.

Coving required for 15 kitchens = 15×3900 mm

$$= 58\ 500 \text{ mm}$$

Checking procedure: $3900 \times 15 = 58\ 500$

No. of lengths = $58\ 500 \div 2700$

$$= 21.6666667$$

$$= 22 \text{ rounding up to the nearest whole length}$$

Reminder

A useful checking procedure for any division sum is to multiply the answer by the number which you have divided by. The check should be the number you first entered to be divided. See Module 13.

When Susmita does the check this time, she does the calculation the same as Joe and then multiplies the answer on her display by 2700 and gets the starting number 58 500.

If 10% is added for wastage 24.2 length required. Joe and Susmita decide to round this to 25 lengths.

Length of plinth required for 15 kitchens = 15×5770 mm

$$= 86\ 550 \text{ mm}$$

No. of lengths of plinth required = $86\ 550 \div 2700$

$$= 32.055555556$$

$$= 33 \text{ to the next whole length}$$

Reminder

To calculate a 10% increase multiply the amount by 1.1. For further help see Module 10.

If 10% is added for wastage 36.3 lengths are required, 37 to the next whole length

Length of worktop required for 15 kitchens = 15×5935 mm

$$= 89\ 025 \text{ mm}$$

No. of lengths of worktop required = $89\ 025 \div 2700$

$$= 32.9722222$$

$$= 33 \text{ to the next whole length}$$

If 10% added for wastage this becomes the same as for the plinth, i.e. 37 lengths.

Susmita does the calculations and Joe does the checking procedures doing the calculations in a different order.

$$\text{Cost of 25 lengths of coving} = 25 \times £17.79$$

$$= £444.75$$

$$\text{Cost of 37 lengths of plinth} = 37 \times £61.25$$

$$= £2266.25$$

$$\text{Cost of 37 lengths of worktop} = 37 \times £64.25$$

$$= £2377.25$$

$$\text{Cost of 15 sets of knobs} = 15 \times £17.48$$

$$= £262.20$$

$$\text{Total cost} = £444.75 + £2266.25 + £2377.25 + £262.20$$

$$= £5350.45$$

$$\text{Saving for the customer} = £6443.40 - £5350.45$$

$$= £1092.95$$

Calculating the 40% discount

$$£5350.45 \times 0.6 = £3210.27$$

The cost to the customer will be £3210.27

Practice opportunity

Calculate the cost of the specialist materials required for the kitchen which you have designed. Make sure you take advantage of any discount offers or special offers.

Reminder

An axonometric drawing has the following features: A: all vertical lines of the object are drawn vertical; B: all horizontal lines are drawn at 45° to the horizontal on the paper. If you would like further help with pictorial representation see Module 24.

3.6 Doing a pictorial representation of the kitchen

Joe and Susmita prepare a set of neat drawings for the customer. The plans and wall elevations will be included. Joe does the axonometric projection of the kitchen for the customer to show the full detail of their design.

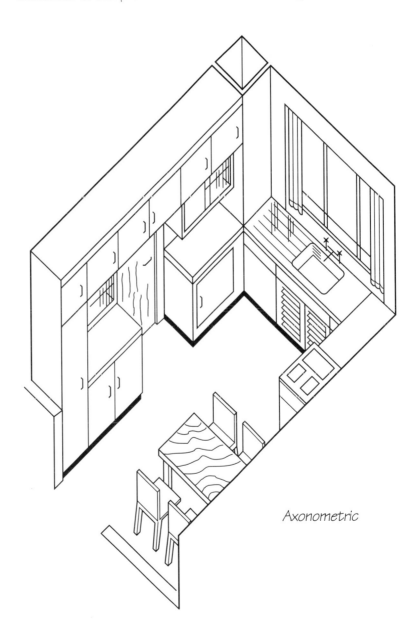

Axonometric

Practice opportunity

Draw a pictorial representation of the kitchen which you have designed. You may like to consider doing an isometric drawing rather than an axonometric one.

3.7 Preparing the estimate

Joe and Susmita prepare the following estimate to hand to their supervisor.

Cost of 15 sets of units @ 3433.58 per set	51503.70
Cost of 25 lengths of coving	444.75
Cost of 37 lengths of plinth	2266.25
Cost of 37 lengths of worktop	2377.25
Cost of 15 sets of knobs	262.20
Sub-total	56854.15
Less 40% discount for bulk purchase	22741.66
Total cost of materials	34112.49

Susmita checks this total which Joe has got by subtracting the discount from the sub-total. She multiplies the sub-total by 0.6 and gets the same answer. *It checks!*

Labour charge for fitting units @ £125 per kitchen	1875.00
Total labour + materials	35987.49
10% Overheads charge	3598.75
Final total costings	39586.24

Again Susmita does a check by multiplying the total costings by 1.1 and gets the same answer.

VAT of 17.5%	6927.59
Total	£46513.83

This last total Susmita checks by multiplying the final costings total by 1.175 and finds that it checks.

Suddenly Joe looks at the estimate and says that the 10% overheads shouldn't have been added before the VAT calculation. Susmita says that it will not make any difference to the final total. Whom do you think is correct?

Practice opportunity

Prepare an estimate for the cost of the kitchen you have designed.

Excavation for a swimming pool

A Number skills

B Job description and analysis of tasks

A local hotel is developing a leisure centre for its guests and the local residents. MartinCrafts have been asked to provide an estimate for excavating and installing a swimming pool. The hotel has already negotiated with the Home Counties Swimming Pool Company with regard to the design and shape of the pool and have given MartinCrafts a drawing on which to base their calculations.

All the dimensions of the pool have been given in feet. The pool is made of heavy duty galvanised steel and is erected on site with an integrated deck support system which measures 3 feet all the way round the perimeter of the pool. Paul and Sally have been asked to do the estimate for the excavation process.

Dimensions

- the swimming pool has a main part which is a rectangle measuring 20 feet by 40 feet
- on the left-hand shorter side there is a semi-circular section which has a diameter of 10 metres and depth of 2 m
- on the right-hand shorter side there is an extra rectangular section which is a flight of steps leading into the main rectangular section of the pool
- the steps go from ground level to the depth of 1 m
- the semi-circular section and the steps are in the middle of the shorter sides
- the depth of the pool is a wedge shape which goes from 1 m to 2 m in the main rectangular section
- the steps have a cross-section view of a right-angled triangle.

Paul and Sally begin by making a list of the tasks they have got to do:

- visit the site and take bearings of the position for the pool
- calculate the volume of earth to be excavated
- calculate how long the digger will have to be hired to complete the excavation
- calculate the cost of hire of the mechanical digger
- estimate how many journeys will have to be made by a lorry to remove the soil
- calculate the cost of the soil removal
- decide on how many lorries should be hired
- prepare an estimate for the customer.

C Cost of materials

The hire of the digger is £75 plus £100 per day.

The hire of the lorries is costed by the number of loads removed to a local land fill site and is currently £42 per load.

D Other considerations

- the size of the bucket on the digger is 0.150 m^3
- the soil has a bulking factor of 20%
- each lorry takes approximately 5 m^3 per load
- a figure of 12.5% is to be added to the cost of the hiring charge for the digger and the lorry to cover overheads
- VAT is to be charged at 17.5%.

4.1 Using ratios and bearings

> ### Reminder
>
> All scaled drawings are described by a unitary ratio. In this example it is the scale of 1:1250. Every measurement on the scaled drawing is 1250 times smaller than it actually is. For more help with scales and ratios see Module 26.

Paul and Sally visit the site and confirm the position of the swimming pool. The block plan they use is drawn in the scale of 1:1250. The actual length of the swimming pool is given as the imperial measure of 40 feet.

Because all the measurements need to be in one system of measurement, Sally uses a conversion constant to change the 40 feet into metres.

She knows that one foot is approximately equal to 0.305 m.

$$40 \text{ feet} = 40 \times 0.305 \text{ m}$$

$$= 12.2 \text{ m}$$

Sally does a quick estimation as a checking procedure for her calculated conversion. One foot is approximately equal to a third of a metre; 40 feet are approximately equal to a third of 40 m, i.e. 40 divided by 3 which is 12 and a bit. Her estimation checks the calculation.

Using the block plan

> ### Reminder
>
> Always remember to check any calculations using some procedure to confirm your answers. See Modules 12, 13, 14.

On the block plan any length is 1/1250 of its actual length.

Paul and Sally use this information to measure the distance of the top left-hand corner of the pool from two different places to locate it. The distance from the top left-hand corner of the pool from the hotel on the block plan is 48 mm due east.

Sally works out the actual distance by multiplying 48 mm by 1250, using her calculator:

$$1250 \times 48 \text{ mm} = 60 \ 000 \text{ mm}$$

Paul checks this by doing a mental calculation. There are eight lots of 1250 in 10 000

Reminder

Conversion factors
are a quick method
of changing a unit
of measurement
into another form.
The conversion
factor to change
metres into
millimetres and vice
versa is 1000. The
conversion factor to
change feet into
metres is 0.305. See
Module 11 for
further help.

$$8 \times 1250 = 10\ 000$$

and there are six lots of eight in 48:

$$6 \times 8 \times 1250 = 6 \times 10\ 000$$

So the answer should be 60 000. *It checks!*

To convert 60 000 mm to metres, Sally divides by the conversion factor of 1000 and gets 60 m. Paul then measures out the 60 m from the rear of the hotel on the bearing of east.

Another suitable place to take a reading from is the top right-hand corner of the hotel's grounds. This again measures approximately 48 mm on the block plan so is also a distance of 60 m.

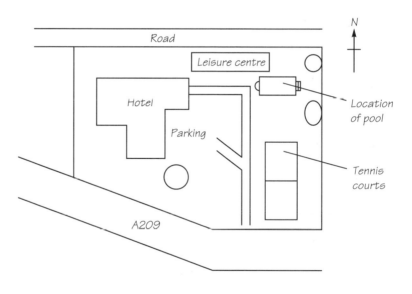

Paul makes notes of what they have measured so far and includes the bearings of the measurements.

The bearing of the top left-hand corner of the pool is to be 60 m due east from the rear of the hotel, and 60 m due 240° from the rear right-hand corner of the hotel's grounds.

Paul and Sally then take two more bearings to locate the top right-hand corner of the pool.

Reminder

Any position can be
located by a bearing
(angle) and a
distance. Two such
readings should
locate a position
exactly. See Module
8 for more help.

Practice opportunity

Either by using a site, block or location plan or otherwise, give two bearings to locate a building or a landmark.

4.2 Calculating the volume of the pool

Reminder

Plans drawn in orthographic projection always include a plan view, a bird's eye view of the object, and elevations which are side and front views of the same object. If you would like more help with understanding drawings see Module 5.

Converting dimensions

Paul and Sally make sketches of both the plan view and side elevation of the pool from the plans, writing in the dimensions.

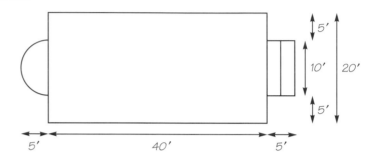

Plan view

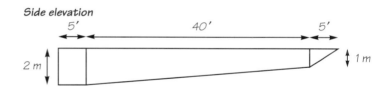

Side elevation

The dimensions of the pool are given in both imperial and metric measurements. The first calculation that Paul and Sally do is to change all the measurements to the metric units.

The conversion factor they use for converting feet to metres is :

1 foot = 0.305 metres

Sally does the following calculations, starting with multiplying by 10:

10 feet = 10×0.305 m

= 3.05 metres

She does all the other calculations using this sum:

Reminder

When multiplying by 10 the place value of each digit in the number moves one position to the left. See Modules 12 and 15 for help.

40 feet = 4 × 10 feet

= 4 × 3.05 metres

= 12.2 metres

20 feet = 2 × 10 feet

= 2 × 3.05 metres

= 6.1 metres

5 feet = ¹/₂ × 10 feet

= ¹/₂ × 3.05 metres

= 1.525 metres

As a checking procedure, Paul uses his calculator and does the following calculations: $0.305 \times 40 = 12.2$; $0.305 \times 10 = 3.05$; $0.305 \times 20 = 6.1$ and finally $0.305 \times 5 = 1.525$. They all check.

Calculating the volumes

The swimming pool is a complex shape and so Paul and Sally decide to divide it up into three easier shaped solid sections.

The main rectangular section

They start by calculating the volume of the main rectangular section in the plan view. On their side elevation, the cross section of this section is a trapezium. Paul draws the following diagram writing in the dimensions they need:

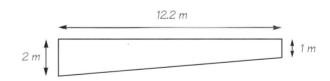

The volume of this section can be calculated by multiplying the area of its cross-section by the width of the pool.

Sally makes notes of what formula they are using:

Area of a trapezium equals the sum of the lengths of the parallel sides divided by two and then multiplied by the perpendicular distance between the parallel sides.

Reminder

It is good practice to write out either in symbols or words any formula which you are using for checking purposes. If you would like further help with using formulas see Module 27. See Module 30 for complex area calculations.

The parallel sides are the depths of the pool at each end, i.e. 1 m and 2 m.

The perpendicular distance between the parallel sides is the length of the pool, i.e. 12.2 m.

> **Reminder**
> _____
>
> Area is always measured in squared units and is calculated using two dimensions; volume is measured in cubic units and is calculated using three dimensions.

$$\text{Area of cross-section} = \frac{1 \text{ m} + 2 \text{ m} \times 12.2 \text{ m}}{2}$$

$$= 1.5 \text{ m} \times 12.2 \text{ m}$$

$$= 18.3 \text{ m}^2$$

The volume of the rectangular section can be calculated by multiplying the area of its trapezium-shaped cross-section by the actual width of the pool.

$$\text{Volume} = \text{area of cross-section} \times \text{width of pool}$$

$$= 18.3 \text{ m}^2 \times 6.1 \text{ m}$$

$$= \mathbf{111.63 \text{ m}^3}$$

Left-hand circular section

The shape of the left-hand circular section is half a cylinder. The area of a cylinder can be calculated by finding the area of its base or its circular cross-section and multiplying by its height.

What is the radius of the circular base? Sally draws a sketch of its elevation:

The radius of the circular cross-section is 5 feet, i.e. 1.525 m. The height of the cylinder is the depth of the pool, 2 m.

$$\text{Volume of cylinder} = \pi r^2 h, \text{ where } r \text{ is the radius and } h \text{ is the height.}$$

Volume of left-hand cross-section = $^1/_2 \times$ volume of cylinder with radius of 1.525 m and height 2 m.

Paul keys in the following calculation into his calculator:

Reminder

Most calculators
have a pi key for
circular calculations.
If yours does not
then you can use
the approximation
of π = 3.142. See
Module 18.

$$\text{Volume} = {}^1/_2 \times \pi \times 1.525 \text{ m} \times 1.525 \text{ m} \times 2 \text{ m}$$

$$= 7.306166415 \text{ m}^3$$

$$= \textbf{7.31 m}^3 \text{ correct to two decimal places}$$

Sally checks this by doing the calculation on her calculator
in reverse order:

$$2 \times 1.525 \times 1.525 \times \pi \times 0.5 = 7.306166415$$

Paul and Sally decide that two decimal places should be
sufficient accuracy for their calculations as there is no need
for exactness.

The flight of steps
The right-hand side of the pool, the flight of steps into the
shallow end has a cross-section of a right-angled triangle.
Paul draws a sketch and writes in the dimensions:

1.525 m

1 m

Reminder

A prism is the name
given to any solid
which has the same
cross section the
whole of its length.
See Module 28 for
further help with
calculating complex
volumes.

The volume of the right-hand section is a prism. The
volume of any prism can be calculated by multiplying the
area of its cross-section by its length.
    Sally makes notes:

Volume = area of cross section × width of the steps.

Area of cross-section = area of a triangle,
i.e. 0.5 × base × height

What is the base of the triangle? It is the depth of the
shallow end of the pool. The height of the triangle is the
horizontal length of the steps.

$$\text{Area of cross-section} = 0.5 \times 1 \text{ m} \times 1.525 \text{ m}$$

$$= 0.7625 \text{ m}^2$$

$$\text{Volume} = 0.7625 \text{ m}^2 \times \text{width of steps}$$

$$= 0.7625 \text{ m}^2 \times 3.05 \text{ m}$$

$$= 2.325625 \text{ m}^3$$

$$= \textbf{2.33 m}^3 \text{ correct to two decimal places}$$

The total volume of earth to be excavated is the sum of the three calculations:

$$\text{Total volume} = \textbf{2.33 m}^3 + \textbf{7.31 m}^3 + \textbf{111.63 m}^3$$

$$= \textbf{121.27 m}^3$$

Practice opportunity

Design a swimming pool shape and calculate the volume of earth which would have to be excavated. Use a circular portion in your design if you would like to practice the area and volume formulas for circular solids.

4.3 How long will the digger be needed?

Having calculated the volume of earth to be excavated, Paul and Sally look at the geological map of the area and confirm that there should not be any unforeseen problems. They have now got to calculate how long the digger will be required.

The calculated volume is 121.27 cubic metres. The digger will not excavate the exact shape but is likely to excavate more than is required. Sally and Paul decide to use a figure for their following calculations which is 10% more than the volume already calculated.

$$\text{Volume of earth to be excavated} = 110\% \text{ of } 121.27 \text{ m}^3$$

$$= 1.1 \times 121.27 \text{ m}^3$$

$$= 133.397 \text{ m}^3$$

> ### Reminder
>
> To calculate a percentage increase of 10% is easily calculated by multiplying the amount by 1.1. If you would like more help with calculating percentages see Module 10.

They decide to work with a volume of 133.4 cubic metres.

How many bucketfuls?

The bucket on the digger which MartinCrafts usually hires is 0.150 m^3. To find out how many times it has got to be filled to remove the earth Paul does the following sum:

$$133.4 \div 0.150 = 889.3333333$$

Sally checks it by doing a multiplication; she multiplies 889 by 0.15 and gets 133.35, which checks as she had left off the decimal fraction part of the number.

They decide to round this up to the next whole number which is 890.

> ### Reminder
>
> When keying in a figure like 0.150 into a calculator, it is important that the decimal point is entered; the zeros at the beginning and ending of the number need not be entered. See Module 4.

Timing the digger

On the plan Paul takes some approximate measurements of the distance that needs to be covered by the digger in each filling and emptying of the bucket. He looks up previous similar jobs' specifications and looks at the approximate times which are taken by a digger doing such distances. He makes a list of 20 of the times observed by the site managers, taken at random:

> 88 secs, 93 secs, 89 secs, 95 secs, 90 secs, 101 secs, 98 secs, 87 secs, 96 secs, 103 secs, 89 secs, 94 secs, 94 secs, 96 secs, 94 secs, 87 secs, 102 secs, 88secs, 97 secs, 95 secs

Reminder

The average that Sally has calculated is called the 'mean'. She could have used other averages such as the mode or median. See Module 20 for further help.

Sally quickly calculates the average time by adding all the times on her calculator and dividing by 20:

$$88+93+89+95+90+101+98+87+96+103$$
$$+89+94+94+96+94+87+102+88+97+95 = 1876$$

$$1876 \div 20 = 93.8$$

$$= 94 \text{ rounded up to the nearest whole second.}$$

They decide to calculate the length of time using 94 secs and 120 secs, i.e. 2 mins per manoeuvre.

Using 94 secs, Paul does the following calculations, checking each answer by doing the multiplication in reverse order and checking the divisions using multiplication:

$$\text{Time needed} = 94 \text{ secs} \times 890$$

$$= 83660 \text{ secs}$$

The number of seconds 83660 is divided by 60 to change it into minutes:

$$= 1394.3333 \text{ mins}$$

Dividing by 60 to change it into hours:

$$= 23.2388889 \text{ hours}$$

$$= 24 \text{ hours to the next whole hour}$$

Using 2 mins, Sally does the following using the same checks as Paul:

Reminder

60 seconds = 1 minute
60 minutes = 1 hour
24 hours = 1 day
7 days = 1 week
4 weeks = 1 month
12 months = 1 year

$$\text{Time needed} = 2 \text{ mins} \times 890$$

$$= 1780 \text{ mins}$$

$$= 29.66666667 \text{ hours (dividing by 60)}$$

$$= 30 \text{ hours to the next whole hour}$$

The working day in fine weather conditions is 8 hours. This means that if each loading takes 94 secs, the digger will be required for 3 days. As they had rounded up approximately 30 seconds to 2 minutes, the second calculation is more realistic and indicates that the digger will be needed for 4 days:

$$30 \div 8 = 3.75$$

This gives them some leeway for bad weather stopping work.

Paul and Sally decide to estimate the hire of the digger for 4 days.

Practice opportunity

Estimate how long the digger will be required to excavate the pool which you have designed. Or for a job with which you are involved calculate the time a particular machine will be needed for hire.

4.4 Cost of hire of the digger

The hire of the digger has a basic charge of £75 and then £100 hire charge per day. Here are Sally's notes:

$$\text{Cost of hire} = £75 + 4 \times £100$$

$$= £475$$

The hire company will charge VAT of 17.5%:

$$\text{Cost of hire} = £475$$

$$\text{VAT} = £83.125$$

$$\text{Total cost of hire} = £475 + £83.125$$

$$= £558.125$$

$$= £558.13 \text{ to the nearest penny}$$

Reminder

To calculate VAT of 17.5% simply multiply the amount by 1.175. If you require any help with calculating percentage increase see Modules 4 and 10.

Sally hands her notes to Paul and he checks the VAT by doing the following:

$£475 \times 1.175 = £558.125$ which checks.

Practice opportunity

Calculate the cost of hire of the machinery you decided to hire for the previous exercise.

4.5 Number of lorry journeys

The lorries which are used by MartinCrafts have a capacity of 5 cubic metres (5 m^3). Paul and Sally have got to estimate the number of journeys required to remove the earth from the site to a local land fill site which is a 30 minute journey away.

The earth as it is excavated bulks up and increases its volume. The earth which the hotel is built on was examined during the preliminary visit and it was estimated that the bulking factor was approximately 20%.

The volume of earth removed = 133.4 m^3

If it bulks by 20%, this constitutes a 20% increase. Sally calculates this in the following way:

Volume of bulked earth = 1.2×133.4 m^3

$= 160.08$ m^3

Paul and Sally decide to work with 160 cubic metres.

If each load taken the site is 5 m^3, the number of journeys needed to complete the job can be calculated by:

$160 \div 5 = 32$

Paul checked this calculation by multiplying 5×32 and getting 160.

The total number of journeys will be 32.

Practice opportunity

Calculate the number of journeys required to remove the earth which has to be removed from your swimming pool or any other similar excavation.

4.6 Cost of earth removal

The local haulage company has given an estimate for the job of removing the earth by costing each journey at £35.

The number of journeys estimated = 32. Paul calculates this using his calculator and Sally checks the calculation by doing it in the reverse order.

Cost of journeys = $32 \times £35$

= £1120

The company will charge MartinCrafts 17.5% VAT. Sally writes out her calculations.

VAT on £1120 = £196

Total cost = £1120 + £196

= £1316

She gets Paul to check her calculation by doing 1120×1.175 and finds that it checks.

Practice opportunity

Research how the removal of earth from a site is costed and calculate the charge which should be made to a client.

5 Material requirements for a detached garage

B Job description and analysis of tasks

MartinCrafts have been asked to build a detached garage from concrete blocks and bricks with a flat roof. The garage is to be situated to the right side of the house and its doors are to be level with the front elevation of the existing property. The customer has already chosen the doors from a Henderson catalogue. Jeff and Sadrin have been asked to make a list of the material requirements for the bill of quantities to be given to the customer with the contract.

After looking at the plans and elevations of the garage and the building specifications, Jeff and Sadrin make a list of the tasks which they have got to do to get the information required for the bill of quantities:

- calculate the materials required for laying the foundations and the floor of the garage
- calculate how many concrete blocks and bricks are required for the construction of the walls
- estimate the mortar requirements
- calculate the materials required for the flat roof construction
- prepare a materials requisition for the building manager which includes the other materials required to complete the garage, e.g. garage doors, lintels, damp proofing.

C Cost of materials

Henderson Garage doors 2134 mm by 1981 mm: £350.

D Other considerations

- 120 bricks needed per 1 m^2 to build a wall one brick thick, i.e. 215 mm
- 10 standard size blocks (440 mm × 215 mm on face) needed for 1 m^2 of walling
- 5% wastage to be added for bricks and blocks
- foundation concrete to be in the ratio of 1:2:4, one cement, two sand and four of 20 mm aggregate
- the floor concrete to be in the ratio of 1:3:6, one cement, three sand and six of ballast
- the specifications state the joists overlap the walls by 200 mm
- roof joists to be inclined 1:40 slope
- joists overlapped on middle beam by 300 mm minimum and securely fixed together
- dimensions of middle beam to be required length × 100 mm × 210 mm
- dimensions of lateral joists to be required lengths × 50 mm × 150 mm
- three layers of roofing felt with top layer being green mineral felt

- roofing boards to be a minimum of 18 mm thick
- 8% wastage to be added to cubic quantity of roof timbers
- 10% wastage to be added to quantity of fascia board and roofing boards
- damp proof membrane on top of hardcore and sand layer
- damp proof in walls 100 mm above the ground.

5.1 Materials required for foundations and floor

Jeff and Sadrin make a rough sketch of the garage plan, writing in the dimensions:

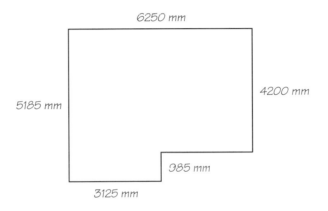

Jeff and Sadrin decide to take the dimensions of the garage as the centre line of the foundations trench. They know that all calculations can only be approximations when dealing with the excavation of earth. The specifications ask for a trench 1 m deep and 300 mm wide. Jeff does the following calculation:

Perimeter of garage = 6250 mm + 4200 mm

+ 5185 mm + 3125 mm

+ 985 mm + 3125 mm

= 22 870 mm

Sadrin does a checking procedure, doing the addition in reverse order:

$$3125 + 985 + 3125 + 5185 + 4200 + 6250 = 22870$$

It checks!

Volume of the trenches = perimeter × width × depth

$$= 22870 \text{ mm} \times 300 \text{ mm} \times 1 \text{ m}$$

Jeff changes all the measurements into metres by dividing the millimetre measurements by 1000:

$$22870 \div 1000 = 22.87$$

$$300 \div 1000 = 0.3$$

Volume of trenches = 22.87 m × 0.3 m × 1 m

$$= 6.861 \text{ m}^3$$

$$= \textbf{7 m}^3 \text{ to the nearest whole cubic metre}$$

To check this multiplication, Sadrin enters the quantities in a different order into his calculator:

$$1 \times 0.3 \times 22.87 = 6.861 \text{ It checks!}$$

Jeff and Sadrin, after checking this multiplication, look at the specifications and see that the foundation concrete requires to be in the proportion of 1:2:4. Adding all the parts of the ratio mix, they get seven parts altogether. This makes the requirements very simple to estimate.

This means that the estimated requirement for the trenches is:

7 m^3 of concrete which is to be mixed in the proportion of 1:2:4 (1 m^3 of cement, 2 m^3 of sand and 4 m^3 of 20 mm aggregate).

The garage floor

The garage floor will be formed by the inner perimeter of the garage wall. However, Jeff and Sadrin use the garage dimensions for the outer wall to give an approximation to use for the suppliers who will generally only sell materials by the whole cubic metre. Also MartinCrafts have been asked to quote for the laying of a concrete drive which means that any surplus materials will be used in the next stage of the job.

The area of the garage floor can be divided into two main sections:

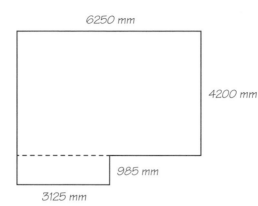

6250 mm

4200 mm

985 mm

3125 mm

Area of larger rectangular section = 6.25 m × 4.2 m

$$= 26.25 \text{ m}^2$$

Area of smaller rectangle = 0.985 m × 3.125 m

$$= 3.078125 \text{ m}^2$$

Total area of garage floor = 26.25 m^2 + 3.078125 m^2

$$= 29.328125 \text{ m}^2$$

$$= 29.33 \text{ m}^2 \text{ to two decimal places}$$

The specifications require 150 mm of hard core with sand to compact as a bottom layer of the floor with a top layer of concrete to a depth of 100 mm.

Volume of hardcore required = 29.33 m^2 × 0.15 m

$$= 4.3995 \text{ m}^3$$

Volume of concrete required = 29.33 m^2 × 0.1 m

$$= 2.933 \text{ m}^3$$

$$= 3 \text{ m}^3 \text{ to the nearest whole cubic metre}$$

Jeff and Sadrin round up the volume of hardcore to 5 m^3 which is one lorry load and the concrete to 3 m^3. The approximations using the outer dimensions have given reasonable quantities.

The concrete mix for the floor has got to be in the ratio of 1:3:6.

The fraction of cement required = $^1/_{10}$ of 3 m^3 = 0.3 m^3

The fraction of sand required = $^3/_{10}$ of 3 m^3 = 0.9 m^3

The fraction of ballast required = $^6/_{10}$ of 3 m^3 = 1.8 m^3

The requirements for the garage floor will be 5 m^3 of hardcore, 3 m^3 of concrete (0.3:0.9:1.8).

Practice opportunity

Make a list of material requirements of a garage of your own design or some building which you can measure and review its plans and elevations.

5.2   Materials required for the walls

Bricks

The plan of the garage showed that the two doors were supported by brick pillars with a connecting brick wall. Jeff drew a sketch of the detail of the front section of the plan, including the information he wanted to use:

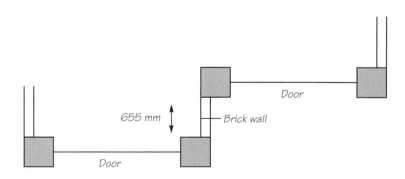

Each brick pillar was 330 mm square, with each layer being made from 4$^1/_2$ bricks:

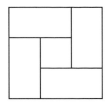

The height of the garage was 2.280 m. Jeff converted this into millimetres and calculated the number of courses of bricks.

The height of a brick + mortar lies between 70 and 80 mm. Jeff decided to use the approximation of 75 mm. So

$$\text{number of courses of bricks} = 2280 \text{ mm} \div 75 \text{ mm}$$

$$= 30.4$$

Sadrin checked this by doing an estimation; he rounded 30.4 down to the whole number 30 and multiplied it mentally by 75.

First he multiplied 70 by 3 and got 210;

then he multiplied 5 by 3 and got 15;

he added this to the 210 and got 225;

all he had to do was to multiply this by 10 and then it would be the same as multiplying by 30:

$30 \times 75 = 2250$. It checked as the 0.4 could account for the extra 30 mm.

Using 31 courses per pillar:

$$\text{Number of bricks} = 31 \times 4.5$$

$$= 139.5 \text{ per pillar}$$

$$= 140 \text{ to the whole brick}$$

$$\text{Number of bricks in four pillars} = 4 \times 140$$

$$= 560$$

The area of the wall joining the pillars in line

$$= 0.655 \text{ m} \times 2.28 \text{ m}$$

$$= 1.4934 \text{ m}^2$$

$$= 1.5 \text{ m}^2 \text{ to one decimal place}$$

Number of bricks per 1 m^2 of 215 mm thick wall is approximately 120.

Number of bricks required for wall

$$= 120 \times 1.5$$

$$= 180$$

Total number of bricks = 180 + 560

$$= 740$$

A 5% wastage is to be added to the amount estimated.

$$740 \times 1.05 = 777$$

Total number of bricks with 5% wastage added = 777

Blocks

Jeff and Sadrin began by calculating the number of blocks needed to build the wall opposite the garage doors.

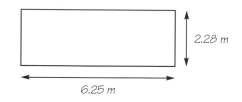

2.28 m

6.25 m

Jeff makes notes of the method he is using:

Area = length × width

$$= 6.25 \text{ m} \times 2.28 \text{ m}$$

$$= 14.25 \text{ m}^2$$

> **Reminder**
>
> $$5\% = \frac{5}{100} = 0.05$$
> To calculate 5% increase, multiply the amount by the decimal 1.05. See Module 10 for help.

> **Reminder**
>
> The area of any rectangle can be found by multiplying its length by its width. See Modules 15 and 16.

Sadrin checks the calculation by doing the multiplication in a different order. He keys in 2.28×6.25 and gets the same answer.

The dimensions of the longer side wall have to be calculated. At one end there is a brick pillar and at the other, the width of the adjoining wall has to be taken into account.

Jeff draws a sketch:

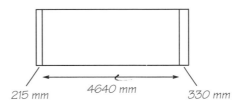

215 mm   4640 mm   330 mm

The brick pillar measures 330 mm and the wall adjoining is 215 mm thick. Jeff subtracts these measurements from the overall length of the wall to calculate the length of wall which remained:

$$5185 \text{ mm} - 215 \text{ mm} - 330 \text{ mm} = 4640 \text{ mm}$$

Sadrin checks by adding 215 and 330 to the answer 4640.

Again, Jeff writes notes to show what he is calculating:

$$\text{Area} = \text{length} \times \text{height}$$

$$= 4.64 \text{ m} \times 2.28 \text{ m}$$

$$= 10.5792 \text{ m}^2$$

Remaining wall

Sadrin draws a diagram showing the dimensions. The brick pillar and the adjoining wall have to be taken into account:

215 mm   3655 mm   330 mm

Sadrin subtracts these from the overall length of the wall and writes them in on the diagram; Jeff checks the calculation:

Reminder

It is good practice to write out in words or symbols any formula which you are using. It provides information for others reading your work and is good for checking that the correct dimensions have been used. See Module 27 for further help.

Reminder

Remember to make sure that all dimensions are given using the same unit of measurement when calculating area or volume.

$$4200 \text{ mm} - 215 \text{ mm} - 330 \text{ mm} = 3655 \text{ mm}$$

$$\text{Area of blocks required} = 3655 \text{ mm} \times 2.28 \text{ m}$$

$$= 3.655 \text{ m} \times 2.28 \text{ m}$$

$$= 8.3334 \text{ m}^2$$

Having calculated all three walls, Jeff and Sadrin calculate the total area using the answers they have got on their calculators.

$$\text{Total area of blocks} = 8.3334 \text{ m}^2 + 10.5792 \text{ m}^2 + 14.25 \text{ m}^2$$

$$= 33.1626 \text{ m}^2$$

Ten blocks are used per 1 metre squared so:

$$\text{Total number of blocks required} = 33.1626 \times 10$$

$$= 331.626$$

$$= 332 \text{ to whole number of blocks}$$

With 5% wastage to be added, it brings the number required to 332×1.05, i.e. 348.6. Rounding this up to the nearest whole 10 is 350.

Number of blocks required = 350

Practice opportunity

For a garage which you have designed or for a building which you can measure, calculate the number of bricks and blocks required for the construction.

5.3 Calculating mortar requirements

Jeff and Sadrin consult a book of tables which they have for reference in their office and get the following information about mortar requirements for the bricks and blocks which they are using. The figure given is only an approximation but is a useful guide:

1 m^2 of 215 mm standard size blocks requires 0.022 m^3 of mortar

1 m^2 of 215 mm thick brick wall requires 0.5 m^3 of mortar

Blockwork

Total area of blockwork $= 33.1626 \text{ m}^2$

$$= 34 \text{ m}^2 \text{ to the next whole metre squared}$$

Mortar requirements $= 34 \times 0.022 \text{ m}^3$

$$\mathbf{= 0.748 \text{ m}^3}$$

Brickwork

The area of the wall joining the pillars $= 1.5 \text{ m}^2$

To calculate the mortar required for the pillars poses a problem because the table only gives approximations for walls. Each pillar is 330 mm square so Jeff and Sadrin decide to think of each pillar as being cut in half and forming a wall 660 mm wide, with a height the same as the garage.

Area of one brick pillar $= 660 \text{ mm} \times 2.28 \text{ m}$

$$= 0.66 \text{ m} \times 2.28 \text{ m}$$

$$- 1.5048 \text{ m}^2$$

Area of four brick pillars $= 4 \times 1.5048 \text{ m}^2$

$$= 6.0192 \text{ m}^2$$

Total brickwork area $= 1.5 \text{ m}^2 + 6.0192 \text{ m}^2$

$$= 7.5192 \text{ m}^2$$

$$= 7.6 \text{ m}^2 \text{ rounding up to the next tenth of } 1 \text{ m}^2$$

Mortar requirements $= 7.6 \times 0.05 \text{ m}^3$

$$\mathbf{= 0.38 \text{ m}^3}$$

$$\text{Total mortar requirements} = 0.748 \text{ m}^3 + 0.38 \text{ m}^3$$

$$= 1.128 \text{ m}^3$$

$$= \mathbf{1.2 \text{ m}^3}$$

Practice opportunity

Calculate the mortar requirements for the garage which you have designed or for another building which you have been working on.

5.4 Materials for the flat roof construction

Roofing timbers

Jeff and Sadrin made a rough sketch of the roof timbers from the plans of the garage:

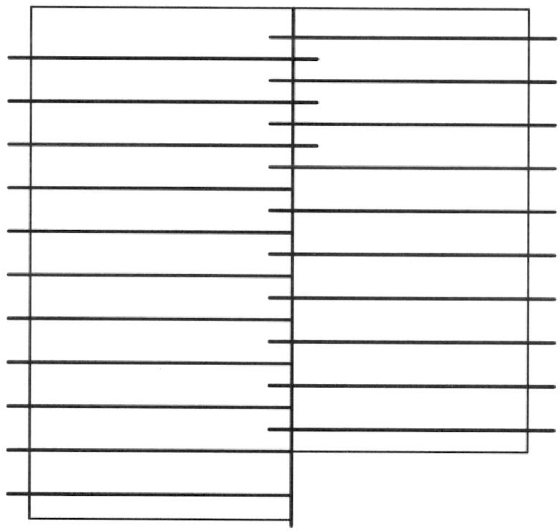

There was one middle beam which ran the full length of the garage and overhang the front by 200 mm.

$$\text{Length of middle beam} = 5185 \text{ mm} + 200 \text{ mm}$$

$$= 5385 \text{ mm}$$

There were 22 lateral joists which had to overlap by 300 mm on the middle beam and overhung the sides of the garage by 200 mm.

Jeff drew an enlarged diagram of two ends of the lateral joists to help with the calculation of the required length:

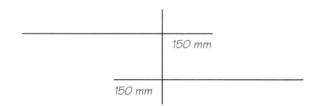

If each joist extended 150 mm over the middle of the middle beam, the resulting overlap would measure 300 mm which was required by the specifications.

Length of each joist = overhang at wall edge
+ middle overlap
+ half the length of the end wall

= 200 mm + 150 mm + 3125 mm

= 3475 mm

Requirements: 1 of 5385 mm × 100 mm × 210 mm

22 of 3475 mm × 50 mm × 150 mm

The timber would have to be given in cubic metres on the bill of quantities.

$$5385 \text{ mm} \times 100 \text{ mm} \times 210 \text{ mm} = 5.385 \text{ m} \times 0.1 \text{ m} \times 0.21 \text{ m}$$

$$= 0.113085 \text{ m}^3$$

$$22 \text{ of } 3475 \text{ mm} \times 50 \text{ mm} \times 150 \text{ mm} = 22 \times 3.475 \text{ m}$$
$$\times 0.05 \text{ m} \times 0.150 \text{ m}$$

$$= 0.573375 \text{ m}^3$$

Total cubic metreage of timber = 0.573375 m³
+ 0.113085 m³

$$= 0.68646 \text{ m}^3$$

A figure of 8% wastage had to be added to this quantity:

$$0.68646 \text{ m}^3 \times 1.08 = 0.7413768 \text{ m}^3$$

$$= \mathbf{0.75 \text{ m}^3} \text{ correct to two decimal places}$$

Roofing boards and felt

The area of the roof was larger than the floor plan due to the overhang. The joists overhang both sides and the front by 200 mm. Sadrin drew a sketch showing the new dimensions:

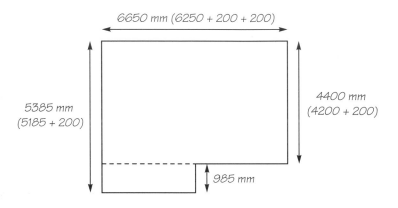

Sadrin calculated the area of the roof by dividing it into two rectangles.

$$\text{Area of large rectangle} = \text{length} \times \text{width}$$

$$= 4.4 \text{ m} \times 6.65 \text{ m}$$

$$= 29.26 \text{ m}^2$$

$$\text{Area of small rectangle} = 3.325 \text{ m} \times 0.985 \text{ m}$$

$$= 3.275125 \text{ m}^2$$

$$\text{Area of roof} = 29.26 \text{ m}^2 + 3.275125 \text{ m}^2$$

$$= 32.535125 \text{ m}^2$$

$$= 33 \text{ m}^2 \text{ to the next whole } 1 \text{ m}^2$$

The dimensions of each roofing board are 2400 mm by 1200 mm.

The area of a roofing board $= 2.4 \text{ m} \times 1.2 \text{ m}$

$$= 2.88 \text{ m}^2$$

Number of roofing boards required $= 33 \div 2.88$

$$= 11.45833333$$

$$= 12$$

Roofing felt is supplied in 10 m^2 rolls.
The first two layers will be 66 m^2 and will require seven rolls.
The top layer will require 33 m^2 and will require four rolls.
The length of fascia required is the perimeter of the roof.

Perimeter of roof $= 6650 \text{ mm} + 4400 \text{ mm} + 3325 \text{ mm}$
$+ 3325 \text{ mm} + 985 \text{ mm} + 5385 \text{ mm}$

$$= 24070 \text{ mm}$$

Requirements of boards and felt are:

12 roofing boards 1200 mm $\times$ 2400 mm

7 rolls of ordinary roofing felt

4 rolls of green mineral chipping felt

25 metres of fascia board

Reminder

The perimeter is the name given to the distance all the way round the outside of any object.

Practice opportunity

Make a list of the materials required to construct the flat roof of the garage you have designed or some other building.

5.5 The bill of quantities

Jeff and Sadrin gathered together all the information they had calculated for the material requirements:

Total number of bricks with 5% wastage added = 777

Number of blocks required = 350

Total mortar requirements = 0.748 m^3 + 0.38 m^3

= 1.128 m^3

= 1.2 m^3

7 m^3 of concrete which is to be mixed in the proportion of 1:2:4 (1 m^3 of cement, 2 m^3 of sand and 4 m^3 of 20 mm aggregate).

The requirements for the garage floor will be 5 m^3 of hardcore, 3 m^3 of concrete (0.3:0.9:1.8).

Total cubic metreage of timber = 0.573375 m^3 + 0.113085 m^3

= 0.68646 m^3

A figure of 8% wastage had to be added to this quantity:

0.68646 m^3 × 1.08 = 0.7413768 m^3

= **0.75 m^3** correct to two decimal places

Requirements of boards and felt are:

12 roofing boards 1200 mm × 2400 mm

7 rolls of ordinary roofing felt

4 rolls of green mineral chipping felt

25 metres of fascia board

On the opposite page is the start of the bill of quantities which they produced based on their work.

Item	Description	Quantity
	Foundations	
	Trench foundation 300 mm by	
	1 m deep, below ground level	
	or as confirmed by LA	
	following site inspections.	
	Foundations concrete 1:2:4	
	20 mm aggregate	
A	Concrete 1:2:4	7 m^3
	Floor	
	150 mm well consolidated	
	hardcore/roadstone; sand	
	blinding;	
	1200 g polythene d.p.m.	
	100 mm well consolidated	
	oversite concrete 1:3:6	
	Water bars to garage doors.	
A	Hardcore	5 m^3
B	Concrete 1:3:6	3 m^3
C	Water bars	2
D	1200 gr polythene	30 m^2

Practice opportunity

Produce a bill of quantities for the garage which you have designed or complete the bill of quantities started by Jeff and Sadrin.

Index